HV
5740
.B75
1998

CLEARI

D1168484

JAN - 1999

Chicago Public Library
Clearing Branch
6423 West 63rd Place
Chicago, Illinois 60638-5005

DYING TO QUIT

Why We Smoke and How We Stop

Janet Brigham

Chicago Public Library
Clearing Branch
6423 West 63rd Place
Chicago, Illinois 60638-5005

JOSEPH HENRY PRESS
Washington, D.C.   1998

JOSEPH HENRY PRESS • 2101 Constitution Avenue, N.W. • Washington, D.C. 20418

The Joseph Henry Press, an imprint of the National Academy Press, was created with the goal of making books on science, technology, and health more widely available to professionals and the public. Joseph Henry was one of the founders of the National Academy of Sciences and a leader of early American science.

Library of Congress Cataloging-in-Publication Data

Brigham, Janet.
    Dying to quit : why we smoke and how we stop / Janet Brigham.
      p. cm.
    Includes bibliographical references and index.
    ISBN 0-309-06409-0 (alk. paper)
    1. Cigarette habit—Psychological aspects. 2. Cigarette habit—Prevention. 3. Tobacco—Physiological effect. I. Title.
HV5740.B75   1998
613.85—dc21   98-17906
                                            CIP

First Printing, May 1998
Second Printing, September 1998

Photographs by Harry Heleotis, New York City.

Copyright 1998 by Janet Brigham. All rights reserved.

Printed in the United States of America.

*This book is dedicated
to those who are dying to quit.*

Contents

Foreword

Many of us in the public health arena have been fighting for decades to help create a worldwide society free of the disease and premature death associated with tobacco use. This effort has forced us to confront not only large, wealthy, deceitful industries, but also the nature of tobacco itself. Despite many public health efforts to promote a tobacco-free life, tobacco remains popular virtually everywhere in the world.

The public health struggle, however, is a mere shadow compared with the inner struggles of tobacco users themselves. They must deal not only with an addiction to nicotine, but with the other powerful and seductive effects of tobacco on their bodies and minds. Make no mistake: this is a struggle to the death. Smoking alone kills more than three million people worldwide every year, and is the direct cause of diseases that affect many millions more.

The numbers are so large that we can hardly envision them. Some 1.1 billion people worldwide use tobacco. This represents about 30 percent of the entire adult population of the world. Yet, as Dr. Janet Brigham writes in this book, we have become so accustomed to smoking as part of our culture, we no longer see it. We are so used to the cigar displays in convenience stores, smokers standing outside doorways on break, and groups of youngsters smoking as they walk down a street, we no longer notice these things.

Yet they affect us all.

Nearly forty years ago, a scientist named Rachel Carson wrote a book called *Silent Spring* that served as a wake-up call about the hazards we humans were posing to our environment. It became the herald for a new movement that had the potential to preserve our world.

Now it is time for us to preserve our lives. *Dying to Quit* is written in the same spirit as *Silent Spring*, and it has a message as alarming but as necessary to hear. *Silent Spring* outlined findings from the science of environmentalism. This book is also a book of science—the science of tobacco's effects on those who use it. It is not a how-to book, not a book about politics. It is a book about how human beings' feelings, thoughts, and lives are affected by nicotine and the many other constituents of tobacco.

Smoking must become a topic we care about; it must become a topic we are aware of, and one we follow with interest and concern. Our worldwide society needs not only legislation and public health initiatives, but also knowledge about the way tobacco affects those who use it. We need to know about tobacco if we are to deal successfully with its influence on the life of each smoker—particularly if we are to help those who want to quit. With that knowledge will come the ability to minimize the disease and death associated with long-term use of tobacco. And this is the most propitious time in history to quit, because of pharmaceuticals and nicotine replacement therapy never before available.

The backbone of our public health efforts has been the dedicated labor of hundreds of scientists and other researchers who have painstakingly learned how tobacco works and why its hold is so powerful. As I write this, we have just become aware of current science-based statistics indicating that smoking and other uses of tobacco are increasing alarmingly quickly among our young people. For many, this will not be surprising news, because we have seen this phenomenon on our streets and in our schools; but for most of us it is not welcome news.

The struggles of society are born anew with each person who comes into the world. Politics and science alone will not win the war that each tobacco user wages when he or she chooses to quit. That war must be fought—and can be won—only within the confines of the soul. This book goes far toward providing both the individual and the society with the weapons needed to succeed.

C. EVERETT KOOP, M.D., Sc.D.
U.S. Surgeon General 1981–1989

On the Runway

Her seat back and tray table were in the upright position. Her seat belt was securely fastened. She couldn't leave her seat, couldn't even recline her seat. Couldn't smoke.

As the plane taxied for take-off, she slipped a pad of yellow legal paper from her handbag and pulled a flight magazine out of the seat pocket in front of her. Using the magazine for a writing surface, she recorded a date and time on the top page of the legal pad. It was late in the morning of a late-summer day; she was flying home for her mother's funeral.

The woman wrote words she probably never expected or wanted another human being to read. She alluded to raising children now grown, to being divorced and remarried. She wrote of sacrificing to build a career, without mentioning what career that was. She wrote too about her mother, about her own daughter, and about her squabbles with both. She scrutinized herself and found herself lacking.

More than anything else, she wrote about smoking, about how it was an illness, a priority, an isolating factor. She wrote that she had used it to put distance between herself and those she loved. As the plane took off, she vowed to set things right by quitting smoking. She resolved to change her life, to feel more, to be kinder, more loving, more patient. She considered quitting before returning from the funeral, then added with ominous innocence that she still had a couple of packs of cigarettes she should finish first.

She apparently folded the notepaper and tucked it in the seat pocket along with the magazine. Later, as she left the plane, she left the pages behind. Maybe she just forgot them; maybe they had already served their purpose.

Not long afterward, an ex-smoker sat in that same seat on another flight. This second woman—my cousin—reached into the seat pocket to look for (no kidding) a barf bag to mail to a friend as a joke. She found, instead, the handwritten pages. She read them with easy comprehension; this was how she felt about smoking, too. Unable to return the pages to their anonymous author, she shared them with a couple of people close to her. She showed them to her husband, who confirmed that he too had felt the same way about smoking. She also sent them to me.

Thus these two women, my cousin, whom I know well, and the airline passenger I may never meet, unknowingly started this book eight years ago. I wish I could share those wrenching paragraphs with all who read this book, but they are not mine to share. They belong to the woman who wrote them. In their stead are the words of another smoker, an articulate and thoughtful young woman who once begged her grandfather to give up his pipe, and now hopes to give up her own cigarettes. The story of her life with and without tobacco forms the framework of this book.

To these three women I owe a sizable debt: Before I read the words from the discarded diary, I regarded smoking mostly in terms of nanograms per milliliter of nicotine metabolites in the bodily fluids of tobacco users. That's what I'd measured, in the quantification of tobacco exposure necessary for studying nicotine and tobacco. My training was in behavioral pharmacology, with occasional doses of brainwave research. I'd had little recent contact with the type of psychological research in which one considers meanings and symbols. I'd had plenty of contact with smokers, but I confess to not always allowing myself to see the whole of their behavior.

Smoking is more than a behavior, and more than a meta-message about our world and our times. Our vulnerability to and widespread acceptance of this practice points us toward the need to explore the complexity of tobacco use, and to search for synthesis in the wealth of scientific research about nicotine and tobacco.

And a wealth it is. Scientists have studied how we puff, how we inhale, how we absorb, how we develop physical tolerance, how nicotine affects our moods, how we start smoking, and how we stop smoking. Little about tobacco use has gone unnoticed, although much remains to be explored. To future historians and scientists a few millennia down the pike, tobacco may appear to be the major bafflement of our time. Even if we survive pollutions, deforestations, wars, rap music, and Windows 95, will enough of us survive our self-made tobacco pandemic? Will future explorers wonder, as many today have questioned—

- How did we come to use tobacco with such frequency and regularity, when we knew that it would kill so many of us?
- How did our need overshadow our wisdom?
- Why didn't we stop?

Around the world, from Sweden to Australia, tobacco scientists spend their working days identifying patterns, contrasts, and logic in tobacco-related biology and behavior. The body of research they generate forms the groundwork for solutions to the worldwide tobacco problem. Although the information in this book is as current as today's news, I note with pleasure that every day this book's contents become increasingly dated. Daily, what researchers call "the field"—the state of the science, based on the most current publications and presentations— moves closer to understanding what keeps the woman on the airplane and her 1.1 billion worldwide counterparts using tobacco. Every day, we move closer to offering them better ways to quit.

Acknowledgments

Anyone writing about tobacco research stands on the shoulders of many giants. I offer grateful thanks to the world's hundreds of nicotine and tobacco scientists, particularly those who reviewed portions of the manuscript and offered suggestions: Neal Benowitz, Warren Bickel, Steve Heishman, Jack Henningfield, David Gilbert, Harry Lando, Ed Lichtenstein, Ken Perkins, and Gary Swan. James Picht's insights also were helpful.

I also offer my gratitude to the anonymous smoker whose story creates the framework for the book.

Additionally, I am thankful that Stephen Mautner of Joseph Henry Press was sufficiently enchanted with the prospectus for this book to encourage me to proceed.

And I can never offer enough warm words about my family and other loved ones for their continuing endurance and good humor.

Janet Brigham
Menlo Park, California
February 1998

DYING TO QUIT

CHAPTER | 1

W e sit alone in her office. She leans back, crosses her legs, and puts me—the interviewer—at ease. A cigarette would look natural in her hand.

"Now, here's an interesting story," she starts. "I go through lots of guilt about this. Neither of my parents smoked. The only person I knew who smoked was my grandfather, and he smoked a pipe.

"When I was—oh, God, in the first grade?—we found out at school how bad smoking is, and all this stuff that somehow you end up not caring about in your teen years. I remember getting down on my knees, crying and begging my grandfather to quit smoking. I told him that it was killing him. Begging him to quit smoking!

"And he did. And here I am smoking, how many years?" She doesn't quite stop to count.

"I don't smoke around him. He probably knows I smoke, but he doesn't say anything, and I don't tell him. I don't smoke before I see him, and I don't smoke until after I *finish* seeing him. Christmas is really difficult, because I can't wait to go have a smoke.

"Smoking was a bad thing, a horrible thing. I do it! I'm a horrible person!" She punctuates her self-condemnation by flinging up well-manicured hands and laughing. She is not a horrible person. She is merely a woman who smokes.

A Drug for All Reasons

Tobacco is everywhere and nowhere. Tobacco is clamped in the mouth of the woman in the car ahead of yours. It dangles from the fingers of the teenager walking toward you. Its stubs pack the crevices of the sidewalk at your bus stop. As you walk down a street, an elderly man walking ahead of you flips away a cigarette with a motion he has probably performed longer than you've been alive. The backpack you buy for your daughter's schoolbooks comes with a zippered cigarette case sewn into the strap. And when you rent a movie you first saw several years ago, now you notice that the good guy is smoking a cigar. You rent another movie and see that this time it's the bad guy. The night you take the garbage out to the curb, you spot a pinprick of light hovering back and forth across the neighbors' porch and realize that the neighbor boy has taken up smoking. Tobacco is everywhere, even where you are so used to seeing it that you no longer notice it.

It is also nearly invisible. Most of us don't know that every hour, around the clock and around the world, 342 people die of tobacco-related disease. One of us dies every ten seconds. Perhaps these numbers go largely unnoticed because tobacco-related disease is a veneer spread across the world, touching every country and every continent, every day. Virtually everywhere, people use tobacco; not all of them die from it. The pandemic seems selective and whimsical. Some smokers consume a pack a day until they die of an unrelated cause in their 90s. Other smokers die unexpectedly of cardiovascular disease in their 30s. Then again, so do some nonsmokers. Scientists can predict that tobacco-related disease will strike a certain proportion of users, although they cannnot predict precisely who those users will be.

Even though about three million smokers die worldwide every year, obviously they don't all die on the same day. When the disease and death are spread out across the months and years, we become inured. Perhaps if it all occurred on one cataclysmic day, in one terrible afternoon, the news media would overflow with stories of grief and waste. Instead, the ultimate and untimely losses from tobacco occur in a hospital room here, a hospice there, a home somewhere else. Scattered tragedies form a picture only when we step back statistically.

Even as the death and disease continue, some pundit criticizes the news media for portraying smokers as victims of the tobacco industry and slams the scientific community for attempting to scare smokers into stopping. It echoes the rebuke from British philosopher Roger Scruton in the *Wall Street Journal:* "It says something about our times that one of the only moral crusade [sic] that has the consistent backing of modern governments is directed against a source of sensual pleasure that does not directly flow, as drugs and pornography do, into the swelling river of delinquency."

Such attacks and counterattacks have become commonplace. On the news, we see that Congress will be considering a hefty hike in tobacco taxes. An hour later, a fictional hero puffs on a cigar. As he contemplates how to save one beautiful woman from going to prison, another sits at his table and orders a drink ("Make it a double"). She ignores his cigar smog; many of us less-beautiful folks who are less used to cigars would be coughing, our eyes puddling, our stomachs churning.

Meanwhile, a CNN Headline News segment features cigar paraphernalia. And the same day, the Internet's Businesswire tells us that 7-Eleven stores, "the first name in convenience," will now be the "first name in premium cigars." More than 3,000 7-Eleven stores in the United States will cater to cigar smokers (the typical one is a 33-year-old college-educated male) by offering high-quality premium cigars. The price will range from $1.50 to $18. That's per cigar. And to keep cigar smokers well equipped, the stores will carry cigar cutters and butane lighters. At Christmastime, the stores will also sell humidors, cigar ashtrays, leather carrying cases, and other gift items. No 33-year-old college-educated male should be without them.

These goods ought to sell well. The chain's cigar market has grown at a rate of 47 percent per year recently.

Observers of popular culture chuckle and wave off concerns about the resurgence of tobacco's popularity. They say the cigar boom and the upsurge in youth smoking, if they are even real, are just fads, no more enduring than hula hoops. In a contrary spirit, those who follow tobacco trends see in the upswing a dangerous relaxation of the awareness of the risks of smoking; the anti-tobacco movement was slow to develop momentum and could fade quickly, they fear.

One tobacco researcher, former government scientist Jack Henningfield, expressed concern over tobacco's upsurge when he addressed a conference of psychologists in 1996. Never, he told the group, had he thought that in the mid-1990s he would be delivering the news he was about to give. Twenty years earlier, smoking was declining in the United States. Now the country's smoking decline had "basically flattened out," and, in fact, smoking appeared to be increasing for the first time in decades. Many of the scientists listening already knew it, but few had expected it. Even when tobacco use had leveled off in the mid-1990s, it was unclear at first whether that represented merely a blip in the overall downward trend, or whether it was the end of the trend.

"Four years in a row of increased [tobacco] use by young Americans," Henningfield announced, adding that the cheaper price of generic cigarettes had made them more accessible to youth. Canada's relaxation of cigarette taxation had increased rates of tobacco use there as well. On top of that, the use of smokeless tobacco, or snuff, had increased dramatically. "The demographics flipped," he explained. In the 1970s, most users of smokeless tobacco were men over 50. Now those men had died, and smokeless tobacco use had shifted to a younger population. The number of new cases and the total number of people affected were increasing.

That wasn't all of the bad news, either. Data from thousands of young smokers indicated that they wanted to quit but had little or no access to effective treatment. In two polls, about 40 percent of young smokers had expressed interest in seeking treatment to quit using tobacco, but treatment was unavailable to them. More than half of them had tried to quit but had failed.

Only a few years earlier, society and scientists alike had believed that tobacco's spell was broken. Few in the health professions would have expected tobacco use to increase following decades of public health campaigns about tobacco-related disease, after limitations on tobacco advertising, after warning labels appeared on cigarette packages, and after smoking was banned in many public places. Epidemiologists were optimistic. However, the decline in tobacco use was neither as permanent nor as pervasive as they had hoped.

In the 1960s and 1970s, smoking still embodied sophistication and finesse. Health professionals didn't treat smokers for nicotine addiction, because smoking wasn't considered an addiction. Even most scientists considered tobacco use a habit, albeit a more dangerous one than popping one's knuckles or chewing one's fingernails. A 1964 report of the U.S. Surgeon General did not apply the word *addiction* to tobacco use because scientific evidence was incomplete. It did, however, indicate that nicotine was the primary reinforcing pharmacologic agent in tobacco that led to continued use. Smoking was tolerated in most public and social settings. As science writer Mark Gold recalled, "A cigar after dinner and a cigarette after sex were culturally embedded in our behavior. Marlboro country was the state we all wanted to live in."

In that climate, a nonsmoker who didn't get a nonsmoking seat on an airplane could end up in the middle of the smoking section. Even though FAA regulations specified that the nonsmoking section should be expanded to meet the needs of nonsmokers, airline crews were reluctant to offend smoking passengers by moving the boundary. Similarly, people often smoked at work. Before the advent of workplace smoking bans, nonsmokers were expected to breathe secondhand smoke without complaint. When a nonsmoker forced to breathe environmental tobacco smoke developed smoke-related health problems, only the most enlightened managers created nonsmoking zones, and these were usually within a few feet of the smoking area.

Many institutions implemented nonsmoking policies in the 1980s and 1990s in response to hundreds of scientific studies exploring the health risks of environmental exposure to tobacco. The government limited tobacco advertising in broadcast media. Cigarette packaging carried health warnings. Municipalities controlled smoking in public places and on public transportation. For a time, tobacco use declined in almost every segment of the population.

Then, with the unpredictable changes of a tight election or a close ball game, tobacco use rose again. The smoking limitations remained, as if society were using them to prove it had paid its dues to health, but the culture began to turn back toward tobacco with the same winking indulgence that had allowed tobacco to become ingrained in society in the first place. This time, though, it was not in innocence or ignorance. This time, most people knew that cigarette filters didn't prevent cancer, and that menthol cigarettes weren't as safe as Ben-Gay. This time, the Surgeon General's warnings were already more prominent than the letters *LS/MFT*. And this time, the United States was helping spread tobacco leaves across the globe.

The U.S.S. *Nicotiana*

An episode of the fictional *Star Trek: The Next Generation* had the starship Enterprise crew encountering two symbiotic human-like groups of beings in an alien planetary system. One group, the Brekkans, produced a substance called felicium that they traded with beings on a neighboring planet. Their interplanetary neighbors, the Ornarans, were survivors of a plague that was cured with felicium. The Ornarans believed that they needed felicium to control their illness, since they sickened whenever they went without felicium.

The two groups' means of transporting goods and medicine between planets had broken down, and the Ornarans, having run out of felicium, begged the Enterprise crew to fix the broken transport. The Enterprise medical staff discovered that felicium was not treating a continuing plague but was actually an addictive drug, which the Brekkans were using to keep the Ornarans locked in a perpetual trade relationship. If the Ornarans ceased trading with the Brekkans, they would not receive their felicium and would experience withdrawal, which they had mistaken for an illness. This is exactly what the Enterprise captain then allowed to happen. He decided that he must follow his fleet's "prime directive" of noninterference and must let the consequences follow a natural outcome. The Ornarans would be forced to break their addiction to felicium. (This, predictably, made them Ornary.)

The episode probably wasn't written specifically to be about the drug nicotine, but it could have been. As with many substances that can lead to addiction or to physical dependence, nicotine is the trap or the hook in tobacco. It is believed to be the primary addictive agent among the thousands of identified compounds in tobacco. It is a major part of what keeps most smokers smoking.

Nicotine is a remarkable drug. Depending on the rapidity and the route of its delivery, it can be a stimulant or it can reduce feelings of anxiety. In high-stress times, it can calm; in down times, it can provide a mild mood elevation. It can enhance thinking and sharpen mental focus. It can help the user relax. Once a person develops physical tolerance to nicotine and overcomes an initial nausea, nicotine becomes a drug for all reasons.

Several recently published scholarly histories of tobacco trace its early use among native tribes in the Americas. The wild form of the tobacco plant *Nicotiana* was spread by natural events over millennia.

Cultivated *Nicotiana*, on the other hand, appears to have been profoundly relevant to the horticultural people who cultivated and dispersed it. Only some of the species of *Nicotiana* produce enough nicotine for use in chewed, sucked, or smoked tobacco.

Old World Europeans, first encountering tobacco in the Americas, were unfamiliar with its use, although many plants other than tobacco had been used for chewing and inhaling throughout the world. In 1492, men from Christopher Columbus' crew visiting the northern coast of Cuba were the first Europeans to witness tobacco smoking. Various explorers recorded that natives used tobacco ceremonially by smoking a type of cigar and blowing tobacco smoke through a long pipe. Amerigo Vespucci may have seen native people chewing tobacco, but the account is unclear as to where Vespucci encountered them or what the native people were chewing. Tobacco was indeed well known to many native peoples along the northern coast of South America. In fact, a form of rolled tobacco served as money. However, more people chewed it or used it as snuff.

Europeans observing and copying the many means of taking tobacco into the body learned all too soon of tobacco's addictive properties. Unlike the native peoples who used it for sacred purposes such as enhancing fertility, predicting weather, conducting war councils, and enabling vision quests, the Europeans used it mainly because they liked using it. It also served as an analgesic for the pain associated with syphilis.

Smoking eventually came to be viewed as sinful, as its use spread throughout Europe and Asia. In some countries, smoking was punishable by mutilation or death. In other countries, tobacco quickly became a valuable crop. Despite prohibitions and condemnations, tobacco use spread rapidly. Tobacco was believed to have medicinal properties and thus was prescribed by physicians for a variety of ailments.

Tobacco was so integral to early colonial culture in North America that the first slaves purchased from Dutch slave traders were bought with tobacco. An early colonist could pay a fine, finance a marriage, or bury a deceased relative by paying in tobacco tender. Historians specu-

We Are the World

30%	adults worldwide who are regular smokers
1.1 billion	number of smokers worldwide
14%	fifth-grade boys in Moscow, Russia, who smoke
53%	tenth-grade boys in Moscow, Russia, who smoke
28%	tenth-grade girls in Moscow, Russia, who smoke
37%	Caribbean men who smoke
60%	men in Spain who smoke
24%	women in Spain who smoke
53%	women physicians in Spain who smoke
73%	Vietnamese men who smoke
24.6	number of cigarettes that one can buy for the price of one McDonald's Big Mac in the U.S.
76.7	number of cigarettes that one can buy for the price of one McDonald's Big Mac in South Korea

late that tobacco taxes and tobacco debts to England contributed to the unrest that led to the Revolutionary War. Tobacco also helped the colonies win that war.

Tobacco use continued to spread throughout the next century and a half, but the virtual explosion in smoking came after the development of machine-rolled cigarettes in the early 1880s. Automated cigarette manufacturing meant that one machine could produce thousands of cigarettes per hour, continuously. As the population of the United

States swelled with immigrants, smoking expanded among the poorer classes. Tobacco taxes came and went with successive wars, but the ranks of smokers steadily increased. Many factors fanned the flames—soldiers given free cigarettes during wartime, children exposed to advertising in print and broadcast media, women assuming the male role of smoker as they assumed jobs previously held only by men. Some 4 percent of Americans' income was spent on tobacco products by the late 1920s. Tobacco seemed to ease the stresses of difficult times; even during the Great Depression, Americans who could afford few other pleasures spent nearly 7 percent of their income on tobacco. Per capita consumption of cigarettes in the United States peaked in 1960 at more than 4,000 cigarettes per person per year. U.S. consumption declined over the next several decades in the face of repeated reports of tobacco's health risks. It has now reached the pre-1950 per capita level of more than 2,500 cigarettes per capita per year—nearly 7 cigarettes per day for every person age 18 and older.

No U.S. state, even among the tobacco states, has ever approached the smoking rates of some countries outside the United States. Although tobacco has been used in many countries for centuries, the development of the machine-manufactured cigarette made cigarettes relatively inexpensive and available. Even in areas of the world where tobacco use can be traced to religious practices several hundred years old, the most widespread form of tobacco in use is the modern cigarette.

Also, cigarettes marketed outside the United States can differ considerably from their American counterparts. Cigarettes sold in the United States have lower levels of tar, as determined by machine testing, but the overall mortality risk from smoking is the same for the United States and for other countries, and has not changed in the last 30 years. Tobacco use has become widespread in countries that are ill equipped to handle extant health care problems, much less the future health needs of a nation of smokers.

The statistics worldwide are astounding. On planet Earth, some 1.1 billion people smoke. Just less than one-third of all adults in the world smoke regularly. To cite some examples:

- About half of all men and 8 percent of all women in developing countries are smokers.
- Some 800 million smokers, or about three-fourths of the world's smokers, live in developing countries.
- In one rural village in West Java, 84 percent of the men smoke.
- In China, Indonesia, Japan, the Philippines, and the Republic of Korea, 60 to 70 percent of all men smoke.
- In Indonesia, twice as many higher-income women smoke as do lower-income women workers, lending status to smoking.
- About one-third of the junior high school boys in Beijing, China, reported a decade ago that they smoked. Of these, one-fifth said they smoked at home.
- Two-thirds of all Russian males smoke, which is the highest rate among European nations.
- Tobacco use is responsible for one-third of all cancer occurring on the Indian subcontinent.
- When tobacco consumption falls in developed countries such as the United States, tobacco marketing strategies shift toward developing countries, and consumption outside the United States increases. However, consumption of tobacco in developing countries does not cause U.S. tobacco consumption to decrease; both can rise simultaneously.

The World Health Organization reported: "If current trends continue, the chief uncertainty about this alarming prediction is not *whether* there will be 10 million deaths a year from tobacco, but precisely when, during the early part of the next century." Those deaths will not only occur in old age, but will start when smokers are about age 35. Half of those who die from smoking-related causes will die in middle age, each one losing about 25 years of life expectancy.

The rise and fall of smoking rates has intertwined with public sentiment in the United States. An anti-tobacco crusade in the late 19th century brought about a ban on cigarette sales in 15 states. One of the legislative measures designed to counteract the anti-cigarette movement

Not So Fine, China

65% - 68%	men in Hunan, Helongjiang, and Jiangsu (China) who smoke
up to 21%	women in Hunan, Helongjiang, and Jiangsu (China) who smoke
one-half	approximate amount of the world's tobacco that is produced in China
one-third	approximate amount of the world's cigarettes that are manufactured in China
one-half	proportion of the world's 1976-1986 increase in tobacco use that occurred in China
Chinese government	manufacturer and seller of the majority of cigarettes in China

was the exclusion of tobacco from federal regulation under the Pure Food and Drug Act of 1906. With this precedent, the regulation of nicotine in tobacco remained outside Food and Drug Administration (FDA) regulation until the FDA recently asserted jurisdiction over tobacco.

As tobacco has circled the world, the smoking and anti-smoking groups have circled the wagons. At the center of the antagonism is an ongoing debate about the health risks associated with tobacco use. As smokers have sued tobacco companies and state and national officials have tried to maneuver manufacturers into vast payment agreements and marketing restrictions, the public war over tobacco has made headlines. In the meantime, each smoker's private war has continued unabated.

If tobacco killed everyone who used it, or if the disease process were shorter, the debate might not be so polarized. Instead, half of the people who smoke or otherwise use tobacco do eventually die from other causes. Even so, half of all tobacco users and a proportion of those involuntarily exposed to tobacco smoke eventually suffer disease and death as a direct result of tobacco use.

A series of publications from the office of the U.S. Surgeon General entitled *The Health Consequences of Smoking* chronicled the risks. The subtitles tell the story:

- *The Health Consequences of Smoking for Women* (1980).
- *The Health Consequences of Smoking: Cancer* (1982).
- *The Health Consequences of Smoking: Cardiovascular Disease* (1983).
- *The Health Consequences of Smoking: Chronic Obstructive Lung Disease* (1984).
- *The Health Consequences of Involuntary Smoking* (1986).
- *The Health Consequences of Using Smokeless Tobacco* (1986).
- *The Health Consequences of Smoking: Nicotine Addiction* (1988).

None of these had the societal impact of the 1964 report *Smoking and Health*, an advisory committee report to the Surgeon General that marked the government's first substantive public stance on the health consequences of tobacco use. The 1964 publication concluded that several forms of cancer were caused by cigarette smoking, among them lung cancer and laryngeal cancer in men, and probably lung cancer in women. The report proposed that the risks for emphysema and cardiovascular disease were augmented by cigarette smoking, although the evidence for a cause-and-effect relationship wasn't yet clear.

As a result of this report and the publication of hundreds of scientific studies, the general public learned of the increased risk for lung cancer and emphysema among smokers. The succeeding volumes in the series offered thousands of pages of technical reviews of a sizable body of research literature outlining the multifaceted risks of tobacco exposure.

Big Problems in a Small World

17 million	annual worldwide death rate from infectious disease
19 million	annual worldwide death rate from noncommunicable disease
3 million	annual worldwide death rate from tobacco-related health problems
10 million	estimated annual death rate from tobacco-related health problems by years 2020 to 2030
2 million	estimated annual death rate of Chinese men from tobacco-related disease by the year 2025
600 – 700 million	estimated number of children who will become regular smokers, of those now living in less developed countries
200 – 300 million	estimated number of children who will die someday from smoking, of those now living in less developed countries

The studies documented in these reports were conducted with considerable scientific rigor. For example, to study the benefits of quitting smoking, a researcher must first document the smoker's actual smoking status. This could be determined by biologically measuring the levels of carbon monoxide in the lungs and the amount of nicotine and its metabolites (substances resulting from nicotine metabolism) in body fluids such as saliva, urine, or blood. The extent of a smoker's tobacco use would be measured not only with questionnaires, but also by biological verification from body fluids, and, in some circumstances, even by count-

Not Only That, But the Surfing's Great

one-fourth	proportion of U.S. adults who smoke
20	average number of cigarettes smoked per day by a typical U.S. smoker
20	number of cigarettes per typical pack
more than 95%	tobacco consumed as cigarettes
Kentucky	state with highest rate of adult smokers (28%)
Indiana	state with second highest rate of adult smokers (27%)
Utah	state with lowest rate of adult smokers (13%)
California	state with second lowest rate of adult smokers (16%)
Hawaii	state with highest overall health rating, third lowest in smoking (18%)

ing, weighing, and measuring a smoker's cigarette butts. Such verification is vital in tobacco research, since the risk of tobacco-related disease is directly related to the amount of tobacco exposure.

Despite the rigor and the quantity of the research, some effects of exposure to tobacco remain obscure except to health professionals. For example, many people are unaware that heart attacks rival lung cancer as a leading cause of premature death in smokers. Although the overall effect is somewhat diminished in those who absorb nicotine without smoking it, as would be the case with smokeless tobacco, it is possible that the cardiovascular system is still stressed by nicotine even in those instances.

Many of the adverse health outcomes of smoking are irreversible. Even so, the title of the 1990 Surgeon General's report on tobacco use assumed an optimistic tone: *The Health Benefits of Smoking Cessation*. This volume documented the ways in which quitting can help most smokers, particularly those who have had cardiovascular damage. We now know that quitting smoking at any age, after any period of tobacco use, substantially reduces the risk of premature death. Quitting reduces the likelihood of disease, reduces symptoms from existing disease, and improves the prognosis for some diseases that may have already developed.

A dramatic decline in tobacco use followed the publication of the Surgeon General's series of reports. Many public and private institutions instituted smoking policies as states and municipalities adopted measures to limit involuntary exposure to tobacco smoke. For instance, a cruise line designated one of its liners to be smoke-free. Most hospitals became smoke-free even before regulations forced them to do so. By presidential decree, many federal workplaces became smoke-free. Many restaurants went smoke-free, and those that didn't often faced stiff regulations regarding ventilation of smoking areas. Before the relatively recent smoking restrictions on domestic United States air travel, passengers on commercial airplanes could never completely escape the cigarette smoke circulating throughout the plane. Even many major airports went largely smoke-free, confining the smoking public to semi-enclosed cells with separate ventilation systems, or allowing smoking only in bars.

Ads and Fads

No amount of separate ventilation, no isolation of smokers from nonsmokers, could have isolated a prospective smoker from tobacco advertising prior to the partial bans that followed the Surgeon General's reports. Tobacco advertising was so common that we all could hum the advertising jingles. Long before Joe Camel came and went, the Marlboro man was an international icon. We knew that "Winston tastes good," and that a dedicated Camel man would walk a mile for a smoke. Then those images disappeared from television and radio, and cigarette promotions no longer cluttered such unlikely places as the covers of new high school textbooks.

Smoking, however, was still visible. Not only was smoking common in television and films, but it actually became overrepresented. A disproportionately large percentage of characters in television and films were depicted as smoking, compared to the actual rates of smoking in the United States. At the same time, news media accounts of smoking underrepresented the health risks of smoking, overplaying instead the prevalence of illness and death from less common causes such as drug overdoses.

On television and in movies, tobacco still is presented as a normative behavior. Researchers Anna Hazan and Stanton Glantz at the University of California at San Francisco sampled three weeks of prime-time television programming on the major networks in fall 1992. They coded all "tobacco events," including anti-smoking messages, as they analyzed 157 programs. One-fourth of the programs contained at least one tobacco event. Overall, about one tobacco event occurred per hour of television. Dramas contained more tobacco events than comedies. The researchers determined that more than 90 percent of the depicted tobacco events were pro-tobacco. Men performed three times as many smoking acts as did women. Whites engaged in nearly 80 percent of the events. Most smokers were shown as middle class or rich, with two-thirds working in professional or technical occupations. More good guys than bad guys smoked. Little of this smoke-filled fictional universe matched the actual smoking demographics for the groups portrayed.

Hazan and Glantz also compared the television findings with a previous study they had published about tobacco use in popular films. An analysis of 62 randomly selected films from 1960 through 1990 revealed that more than one-third of the film intervals studied contained a reference to tobacco. Over the 30 years of film, the presence of tobacco was depicted with little change. The depiction of the motivation of the smokers did, however, shift with time, with relaxation a more prominent motive in the 1960s and 1980s. Major characters in the films were depicted less and less as smokers. Although most smoking was done by

The Smoky Screen

10 – 15 minutes	time between film depictions of tobacco use, 1970s
3 – 5 minutes	time between film depictions of tobacco use, 1990s
1.20	smoking depictions per hour of prime-time TV drama
92%	proportion of those depictions deemed pro-tobacco
55%	tobacco-using TV characters depicted as "good guys"
42%	middle-class TV characters depicted as smokers
65%	male movie characters depicted as smokers
relaxation	movie characters' apparent motivation for smoking
doubled	depiction of movie tobacco events involving relatively young characters, from 1960s to 1980s
halved	depiction of movie tobacco events involving older characters, from 1960s to 1980s
business activity	most popular movie smoking context
nearly three times	extent to which smoking among elite movie characters exceeds its actual prevalence in the population
90%	proportion of movies depicting tobacco use, 1960-1991
rarely	how often moviemakers disclose what tobacco companies pay for product placement

(Hazan and Glantz, 1995; Hazan, Lipton, and Glantz, 1994; Stockwell and Glantz, 1998.)

white characters, a small but increasing amount was done by African-American characters. Female smoking increased as well. The most dramatic increase, however, was among young smokers: Tobacco events involving characters ages 18-29 more than doubled, while smoking in the 30-45 age group was almost halved. Even though smoking among elite characters declined over the three decades studied, it remained nearly three times as prevalent as it was in the actual population.

As the prevalence of smoking has been overrepresented in fiction, so its risks have been underrepresented in the various public media. Using a process called *content analysis*, researcher Karen Frost and colleagues examined such news and general readership staples as *Time*, *Family Circle*, *Reader's Digest*, and *USA Today*. They reported in 1997 that "substantial disparities" existed between actual causes of death and the amount of coverage given to those causes by print media. The most underrepresented causes were tobacco use, cerebrovascular disease, and heart disease. Overrepresented causes of death included illicit use of drugs, which received more than 17 times the news coverage that would have been proportional to its actual occurrence; motor vehicles, nearly 13 times; toxic agents, nearly 11 times; and homicide, more than 7 times.

Tobacco received only 23 percent of the expected copy. Nor was it alone in inaccurate representation: The number two health risk factor, relating to diet and physical activity, got the same amount of coverage as illicit drug use, which was the lowest ranking risk factor among causes of mortality. The authors termed these statistics "impressively dispro-portionate."

Why would this occur? The authors attributed it to competition for viewers and for advertising. They cited the tobacco companies' sizable advertising budgets as influencing news coverage. As they noted, "News reporting is also driven by rarity, novelty, commercial vitality, and drama." Their findings were consistent with a 1992 *New England Journal of Medicine* article by Ken Warner and colleagues of the University of Michigan, revealing that magazines carrying cigarette advertising were less likely to publish information about the hazards of smoking than were publications that restricted tobacco advertising.

Edith Balbach, working with Glantz at UCSF, found that money also apparently talks in unlikely places. They compared tobacco-related articles from two widely distributed elementary-school publications, *Weekly Reader* and *Scholastic News*. An event prompting their inquiry was the publication in 1994 of a fifth-grade edition *Weekly Reader* cover story titled "Do Cigarettes Have a Future?" Surprisingly, the article promoted smokers' rights. Controversy centered not only on the article itself, but also on the fact that the company that purchased *Weekly Reader* in 1991 had been the majority owner of RJR Nabisco, the second largest manufacturer of cigarettes in the U.S.

The UCSF researchers tested this question: Was *Weekly Reader* less likely than *Scholastic News*, which was family owned and not connected to tobacco interests, to mention the consequences of smoking, or to discourage tobacco use? The answer, based on the evaluation of six years of both publications, was a definite yes. *Weekly Reader* was more than twice as likely to give a tobacco industry position. *Scholastic News* was significantly more likely to include a clear no-use message. The authors recommended that health professionals monitor the information carried in both publications, which reach a combined 1 to 2 million students per grade level every week.

None other than Joe Camel appeared in full color on the cover of the February 12, 1992, sixth-grade edition of *Weekly Reader*. Joe showed up in *Weekly Reader* eight times, including once as a centerfold, in the publication's issues evaluated by the research team. Joe's presence there reflected the 1991 findings of Paul M. Fisher and colleagues that some 30 percent of three-year-olds and 91 percent of six-year-olds could identify Joe Camel as a symbol of smoking. It may be no accident that the three brands of cigarettes most smoked by young smokers are also the three most heavily advertised brands. Adults, in contrast, are more likely to choose a generic brand that is advertised less and costs less.

But advertising is not the entire story. It is not even the major part of the story. Over the last several years, tobacco companies in the U.S. have spent far more on what are called "promotions" than they have

Light My Ire: Tobacco Titles

"All Fired Up over Smoking" (*Time*, 18 April 1988)

"Butt of Course: In Amtrak's Smoking Lounge, Fresh Air" (*Washington Post*, 6 July 1997)

"C'mon Baby, Light My Fire" (*Time*, 27 January 1997)

"Cigar's Glow Lighting Up Retail Scene" (*The New York Times*, 23 June 1997)

Cigarette Confidential: The Unfiltered Truth about the Ultimate Addiction (1996 book)

Drag (vocal album, k.d. lang, released 1997)

"First Lady Has a Nicotine Fit" (*Toronto Sun*, 31 August 1997)

"Git Along, Little Stogie" (*Business Week*, 22 December 1997)

"Hard Times at the Hard-Sell Café" (*Houston Business Journal*, 7 July 1997)

"Holy Smokes" (*Time*, 20 January 1997)

"Listening to Nicotine" (*Psychological Science*, May 1997)

Smokescreen (1996 book)

"Smoking Gun: A Cigarette Maker Finally Admits..." (*Time*, 31 March 1997)

"State's Harsh Tobacco 'Fix'" (Associated Press, 5 July 1997)

The Cigarette Papers (1996 book)

"The Cigarette Papers: How the Industry is Trying to Smoke Us All" (*The Nation*, 1 January 1996)

"The Noxious Weed That Built a Nation" (*Washington Post*, 14 May 1997)

Dying to Quit (1998 book you are reading)

spent on conventional advertising such as magazine ads and billboards. Promotional items include sporting goods, accessories, caps, and even a wardrobe. Half of adolescents who smoke own at least one promotional item. Many are acquired through attendance at a tobacco-sponsored sporting event. (See table, p. 45.)

In some cases, the promotion is not just an item, but an ambience. A recent promotional approach geared toward young adults has aggressively and successfully pursued the young adult market. In one midwestern city, Camel (owned by RJR Nabisco Holdings, Inc.) has invested money in about two dozen bars popular among young adults. In exchange for exclusively promoting Camel cigarettes, the bars receive Camel napkins, matches, ashtrays, T-shirts, lighters, and display cases. Employees get free Camel cigarettes. Display cases, which hold only varieties of Camels, sit prominently behind the bar. Camel representatives swap a full pack of Camels with customers smoking non-Camel brands.

Camel runs full-page ads in newsweeklies to advertise the bars it has signed onto this promotional program. An ad is called a "Camel Page." Highlighting more than a dozen bars a week, it lists themes, specials, and schedules for Camel nights. Camel pays for the ads, saving each participating bar owner perhaps thousands of dollars a year. In return, Camel has an exclusive association with bars where young adults congregate to drink and smoke. The bars sell only Camels, from a veneer-and-glass display case with a backlit "Kamel" sign on top. Since vending machines could be outlawed in pending tobacco negotiations and legislation, this arrangement positions itself on the side of future legality by selling Camels from behind the bar. Eventually, that might be the only place cigarettes can be sold in bars. Some experts say that this is the tobacco industry's smartest marketing strategy yet, and perhaps the only type of promotion that would not be affected by legislation or industry settlements.

Strategic marketing success is not new to tobacco. Auburn University marketing professor Herb Rotfeld lamented such success in a 1996 commentary for the American Marketing Association's *Marketing*

News. Rotfeld criticized the "misplaced marketing" through which public information might be distorted to the detriment of society. Firms should be allowed to sell their products, he wrote, but marketing raises the question of whether firms should be able to maximize their profits with products that negatively impact public health. "Tobacco firms are efficient marketers," his headline read, adding the provocative question, "Should they be?"

Whatever ground tobacco interests lost in the public arena because of public concern over health risks, they have regained in many private lives. The United States and some other countries take smoking seriously enough to ban it on airplanes, in restaurants, at workplaces, and in public buildings, yet many people are unwilling or unable to ban it in their own homes and vehicles. Like cultures before ours that accepted death from infectious disease as part of the human condition, we have come to accept the risk of death and disease from tobacco use as the consequence of our free agency.

Those who would be the Pasteurs, van Leeuwenhoeks, and Clara Bartons of this preventable cause of human death sometimes find themselves labeled as tyrants trying to abridge human freedom. An examination of the smokers'-rights literature shows that the sides have become deeply polarized. A 1995 report by Teresa Cardador, Hazan, and Glantz analyzed the thematic content of smokers' rights publications. The team found that the literature reflected the tobacco-control stance as a threat to individual rights and free choice. The publications sought to undermine the opposition by refuting scientific evidence related to the health hazards of secondhand (environmental) tobacco smoke. By creating legitimacy for the tobacco industry's position, the literature encouraged readers to perceive the pro-smoking side as "targets of unfair discrimination." Readers learned of political and social action movements that purportedly threatened or supported smokers' rights.

Curiously, the authors then matched the tobacco publications' themes to what is known as the "stages of change" theory, a model of the process of changes in human behavior that has been applied to smoking cessation by Prochaska and DiClementi. The four prevalent themes matched four stages of change, ranging from the *pre-contemplation* phase in which one is on the verge of considering a behavior change, to the stage termed *action*. The authors counted mentions of items in the four themes across a six-year span (1987-1992) and discovered a dramatic increase in mention of social action and efforts to undermine the opposition in the more recent years. The tobacco-control movement was characterized with what the authors termed "contemptuous language," with words such as *hysteria, extremist, smoker-bashing, class hatred, victimization, alarmist, zealous,* and *warfare.* Agencies and individuals involved in tobacco control were portrayed as "liars who ignore the truth and manipulate the public to impose their lifestyle choices on others."

Is the polarization itself changing how we perceive tobacco? As smoking disappears from public view, does its appeal increase? Author Richard Klein considered the relationship: "We are in the midst of one of those periodic moments of repression, when the culture, descended from Puritans, imposes its hysterical visions and enforces its guilty constraints on society, legislating moral judgments under the guise of public health, all the while enlarging the power of surveillance and the reach of censorship to achieve a general restriction of freedom."

As a national climate suppresses public display of tobacco use, does a smoldering cigarette take on an enticing aroma? If smoking were banal, like taking an aspirin or yawning, would famous people bother to be on the cover of *Cigar Aficionado* magazine, or would the magazine even exist? Does our determination to create a healthy culture make unhealthful behaviors appealing?

Answers to such questions do not come quickly or cheaply. The use of tobacco (or any other addictive substance, for that matter) is bewilderingly complex, stretching the definitions and boundaries of both psychopharmacology and behavioral science. It refuses to fit tidily in any one field, demanding instead that addiction scientists also become sociologists, anthropologists, and perhaps even publicists.

Tobacco was once used in sacred ceremonies. When tobacco left the realm of the shaman, as anthropologist Joseph Winter of the University of New Mexico believes, its powers overtook a naïve world unprepared for its hold. Each smoker struggling repeatedly to quit, each grade school child smoking a first cigarette, each of us touched by the life of someone who cannot stop using tobacco—in other words, all of us—lives with the consequence of the widespread profaning of a sacred substance.

Whatever the source of tobacco's control, it affects all who use it. For each of us who uses tobacco, the experience is unique. The sensations are as individual as a voice-print, as stylized as the tilt of an eyebrow, and as irreproducible as a kiss. To use tobacco is to use a drug, to make a statement, to march through a rite of passage, to satisfy a craving. And more.

CHAPTER 2

D o we actually look younger, do our faces lighten up and the years fall away, when we talk about when we were young? For a moment, do we go back before our present problems started, before we made the choices we now have to live with? I wonder, as she reminisces about starting smoking.

"I actually managed to make it through seventh grade without smoking, which was pretty amazing." She shakes her head. "Everybody was smoking, and the reason I didn't do it was because I wasn't part of the in-crowd and I didn't have a lot of friends. It was 'in' to do it. Everybody who was 'in' would smoke in the bathroom. I just wasn't part of the in-crowd. I was a definite outcast, skinny, red-haired, funny-looking . . ."

Skinny, red-haired, and funny-looking now suit her well, I note.

"We moved the summer between seventh and eighth grade, and I ended up going to a private school for kids who had been kicked out of public schools, although my parents didn't realize that when they enrolled me." The school emphasized reading, writing, arithmetic, and languages, she explains. "My parents thought it was a really good, small private school, where I would get lots of individualized attention. Which I did. And I actually did learn a lot better in that environment.

"But unfortunately there was a lot of drug use, and a lot of promiscuity, and nearly everybody smoked.

"I don't actually remember the first time that I had a cigarette. The first time, I don't. It didn't take me long to get over the sickness. I remember that. I remember people saying, 'Oh, no, you'll be sick for a while.'"

She shakes her head firmly from side to side. "Huh-uh, huh-uh. Two or three cigarettes, that's all that made me sick."

And did anyone notice, or try to stop you? I ask.

"There was a teacher at school that I really admired. He smoked, actually, but he tried to get me to quit. There were these little grocery stores that everybody used to hang out at, where there'd be maybe 15 kids there smoking. He saw me there and turned me in, and no one else. I think it's just because he cared about me. I was one of his favorite students. He liked me. And he knew what hell smoking was and didn't want me to get involved in it.

"Mr. North—that's his name. I really liked him a lot.

"I got written up. I didn't get suspended. I was a good kid; I studied hard, I got good grades. I'd never been written up before. I'd never been in any trouble. But I think they told my parents, and that's why my mom tried an experiment on me."

"After I'd been smoking for a while, my mom knew that I'd been dabbling in it, and she wanted to turn me off to it. So she said, 'Well, why don't we get a pack of cigarettes, and we'll smoke together,' thinking that that would discourage me, because then I would see how stupid smoking looked. And she, of course, was hacking and coughing. And I, of course, wasn't."

She transferred to another private school, one where the students called the teachers by their first names. "This was in 1980, when you still didn't do that yet," she explains. "Now you do, ten years later. Before, it was 'Mr. North,' and then one year later I'm calling my teachers 'Bo' and 'Phoebe.' It was a prep school, and we had to dress nicely, but we didn't have to wear uniforms. A lot of rich kids went there.

"There was a smoking area at the school, where a lot of us hung out and smoked. Smoking isn't allowed on campus now, but they used to allow it in that one little spot. It was a social thing."

———

Lighting Up the World

The fictional Douglas Spaulding, a mighty 12 years old, stood in early morning darkness at the window of a cupola at his grandparents' house. He smiled, pointed a finger, and performed his magic. As he willed it, the town began to awaken. He ordered the old people to wake up, the lights to turn on, his grandma and great-grandma to fry hotcakes, birds to sing, and the sun to rise. With a final snap of his fingers, he willed the world awake, and it followed his bidding.

"Yes, sir, he thought"—as author Ray Bradbury composed in young Douglas' voice—"everyone jumps, everyone runs when I yell. It'll be a fine season."

The same adolescent narcissism with which Bradbury's character felt all-powerful and invulnerable in the world offers psychological protection to many adolescents who meander and sometimes screech toward adulthood. It is a feature of many adolescents to believe that they can run and not be weary, drive fast and not crash, stay up all night and be normal the next day, and do dangerous things without consequences. If there are consequences, adolescents typically believe that the consequences are so distant that they don't matter now. Adolescents have greater tasks than worrying about a faraway future: They are becoming someone and becoming separate, a process somehow made simpler by becoming indistinguishable from those around them. They are struggling with relationships, acceptance, hormones, identity, school, religion, death, success, and, all too often, tobacco.

It is a jarring reality that the feeling of invulnerability that can protect them against the slings and arrows of an outrageous world also makes them vulnerable to behaviors that can haunt them throughout their lives. A few minutes of unplanned and unprotected intimacy can become an infant nine months later. A few beers and a brief car ride can become an incarceration for drunk driving. A single cigarette can lead to emphysema or a heart attack in thirty or forty years. Adolescence isn't a time for existing that far ahead. It is a today time. At best a tomorrow or next week time. And because of this, adolescence has become the prime time for tobacco initiation.

Tobacco's initial toxicity seems to provide adolescents with little protection against its continued and perpetual use. No known interventions seem to work consistently either in preventing tobacco use or in stopping it among adolescents. Tobacco prevention specialists find themselves outmaneuvered by the clever wizards behind the tobacco industry's advertising and promotions. Tobacco control countermeasures, however expensive and well planned, all too often backfire.

An example of this is a recent Arizona state initiative to combat teen smoking with T-shirts and other paraphernalia emblazoned with this slogan: "Tobacco. Tumor causing, teeth staining, smelly puking habit." Predictably, many young persons wore the slogan openly as they proudly and defiantly smoked. (As lyricist Tom Jones [*The Fantasticks*] wrote, "Dog's got to bark, a mule's got to bray/. . . Children, I guess, must get their own way/The minute that you say 'no.'")

As the strategies falter, the figures climb. Tobacco use among young people has steadily increased throughout the decade of the '90s. Smoking has increased among teenagers and young adults ages 18-25; smokeless tobacco use, following a boom in the '80s, has not shown the sharp decline many experts predicted; cigar use has become wildly popular and also shows no sign of diminishing.

It is difficult to dismiss these youthful indulgences in view of the statistic that some 90 percent of all smokers begin the practice during adolescence. Every day, about 6,000 U.S. young people try smoking and about 3,000 become regular smokers. Of these, scientists predict that one-third to one-half eventually will die of smoking-related causes. The path from experimentation to becoming a regular smoker usually takes between two and three years. Young persons who take up smoking are likely to keep smoking for up to 20 years.

The statistics on youth use of tobacco are sobering. Even in states with relatively low overall smoking rates among adults, we commonly see groups of young people walking along, laughing, talking, and smoking. Somewhat invisible to us are the sizable numbers of young users of smokeless tobacco, primarily moist oral snuff. Some smokeless users also smoke, but many do not.

Tobacco use is growing faster among adolescents than among any other age group. Even though tobacco use levels off among other age groups, it continues to rise among the young. A 1998 report indicated that more than half of white male U.S. high school students used tobacco in the previous month. Among high school females, the rate was more than 40 percent. Cigarette smoking doubled among African-American high school males from 14 percent in 1991 to 28 percent in 1997. These findings from the U.S. Centers for Disease Control and Prevention (commonly called the CDC), reported results of the 1997 Youth Risk Behavior Survey. About 40 percent of all high school students surveyed in 1997 said they used cigarettes. About 21 percent of high school white males reported using smokeless tobacco, leading all other groups studied. More than 30 percent of all male high school students used cigars.

Recently published data show that urban-rural differences in youth smoking rates have shifted. As researchers Christine E. Cronk and Paul Sarvela of Southern Illinois University explained, "These findings contrast with the popularly held notion that rural youth are more protected against the use and abuse of drugs by their distance from the factors supporting drug use in urban environments." They noted that substance availability has changed in rural areas, that prevention is less common and less effective, and that social factors no longer protect youth from drug use. The risk factors for adolescent substance use are not always obvious or well known. For example, tobacco use is higher among adolescents with learning disabilities than among their non-disabled peers. Researcher John W. Maag at the University of Nebraska-Lincoln and his colleagues speculated that the increased risk for delinquency among those who are learning disabled could contribute to their higher rates of tobacco and marijuana use.

Among homeless and runaway youth, tobacco use is common. About 81 percent of homeless youth living on the streets used tobacco, as reported in a 1997 study from Jody Greene and colleagues of the Research Triangle Institute. Among those living in shelters, the smoking prevalence was 71 percent. Use of other substances, including alcohol and marijuana, was also high.

Nor is adolescent smoking limited to the United States. The usual pattern in a country is for men to take up smoking first, followed by boys, women, and girls, according to Ann Charlton of the International Union Against Cancer (UICC). In many developing countries, cigarette smoking is at the men/boys stage, thus creating what she calls "two different scenarios with regard to young people and smoking." Young persons in Western countries are aware of the health risks of smoking but pay little attention to them. Young persons in developing countries are less aware of potential health risks from tobacco use, and sometimes are receptive to educational programs informing them about the dangers of tobacco.

If smoking rates don't change among today's children, some 30 million Europeans and 50 million Chinese who are now children could die of tobacco-related disease later in life, according to Charlton, who cited World Health Organization estimates.

How do young persons get tobacco, since they can't buy it legally in most places? According to a 1996 Youth Risk Behavior Surveillance

Too Young to Vote

500 million	cigarettes smoked every year by people under age 18 in the United States
16.6 million	U.S. adolescents currently younger than 17 who are likely to become smokers
15	median (middle) age at which smoking starts in the United States
12 – 13	average age at which U.S. females start smoking
5 million	adolescents under age 17 likely to die eventually from smoking-related disease in the United States
more than 3000	children and adolescents who start smoking every day in the United States
about 75%	adolescent smokers in the United States who have made at least one serious attempt to quit smoking
50%	chance that a U.S. adolescent smoker will smoke as an adult
77%	adult U.S. smokers who were daily smokers before age 20
91%	adult U.S. smokers who tried their first cigarette before age 20
20%	high school seniors who smoke daily in the United States
73%	high school senior smokers in the United States who expected not to be smoking in 5 years, but were still smoking 5 to 6 years later

survey published by the CDC, half of high school seniors under age 18 reported simply buying cigarettes at a store or gas station. Some 39 percent reported that they usually bought cigarettes in a store. One-third routinely borrowed cigarettes from someone else, and 16 percent gave someone else money to buy cigarettes for them. Across all high school years, at least three-fourths of the youthful smokers reported not being asked to show proof of their age when they bought cigarettes. Availability varied widely among states: 18 percent of Idaho high school students under age 18 bought cigarettes at a store or gas station, yet 49 percent of New Hampshire students reported such purchases. More than 90 percent of students from some cities, including Washington, D.C., and Detroit, reported not having to show proof of age to purchase cigarettes.

Camels and Caveats

Research into tobacco use, or into any other aspect of human behavior, generally is conducted in one of a several ways. The most typical approach is a *cross-sectional* study, in which a statistically random sample of subjects is chosen from throughout the "population" of the group under study. This would involve, for example, randomly selecting groups of young persons from throughout a particular state or region and surveying their tobacco use by administering questionnaires. The information cited in the preceding paragraphs was collected through use of an 84-item questionnaire filled out by high school students during class time in selected schools throughout the country. The questionnaire could be filled out anonymously, and the researchers got parental consent before administering the questionnaires to the young people.

By its nature, this procedure involves some data collection challenges. Those conducting the research have to ask these questions: Did those filling out the questionnaire tell the truth? Did they understand the questionnaire? Were the respondents sufficiently typical so that their responses could generalize to the population at large? The researchers collecting the data employ sophisticated techniques from the fields of epidemiology and statistics to work around those potential hazards.

Go, Joe, Go

$300	amount Richard Reynolds paid in 1913 to buy the name *Red Kamel* and the last 500 Red Kamel cigarettes from a small New York tobacco company
0.5%	Camel cigarettes' share of the youth market before the Joe Camel campaign
32%	Camel cigarettes' share of the youth market after the Joe Camel campaign
$6 million	Camel sales to those under age 18 before the Joe Camel campaign
$476 million	Camel sales to those under age 18 two years into the Joe Camel campaign
Willy the Penguin	Brown & Williamson's answer to Joe Camel
Mac the Moose	The Maine American Cancer Society's answer to Joe Camel
Red Kamel	A favorite brand among those who were children and teenagers during the Joe Camel campaign

(Epidemiologists, by the way, study not only contagious epidemics such as cholera or measles, but also other conditions within a particular population, such as numbers of deaths by traffic accident, cases of head lice infestation, or occurrence of multiple births.)

Cross-sectional studies can provide useful information about the relationships between factors under study, but they provide little help in determining causality or predicting outcomes. For more powerful predictive analyses that can suggest causation, researchers need to follow people over a period of time. This *longitudinal* research can be expensive to conduct, because of the need for tracking the research subjects and maintaining a research facility. It can be simulated, to some degree, by cross-sectional research that looks back retrospectively to possible precursors to tobacco use, such as school performance or tobacco use within the family where the person grew up. Nonetheless, some aspects of longitudinal research cannot be duplicated through any other approach.

Two words are key in this type of research: *prediction* and *association*. A study that involves prediction (which is actually a precise statistical term, unrelated to prophecy or to a "psychic" phone line) can explain to what extent one factor, such as depression, predicts, but does not necessarily cause, another. In other words, the process provides a numerical indication of how much the presence or absence of a factor can be used statistically to predict the likelihood that another factor will be present. Weather patterns (temperature, humidity, clouds, wind, etc.) can be used to predict whether or not it will snow in Minneapolis or blow in Idaho. As we all know, such prediction isn't a sure thing in weather forecasts, and neither is it in research into human behavior.

As a hypothetical and scientifically unexamined case, consider this example: Imagine that a preference for long baths as a child predicts adult obesity. The baths themselves would not be a cause of obesity, but they may be related to other factors (disinterest in exercise, for example) that are directly related to obesity. A layperson unfamiliar with statistics might see data connecting the two and wonder whether soaking in a tub for a long time in childhood might actually be causing eventual obesity. Once the news spreads, mothers everywhere would be whisking their youngsters through a light lather and a sprinkling of water, hoping to avoid whatever might cause obesity years down the road.

The other key concept, association, also can be misinterpreted. Association is generally demonstrated by some variation or expansion of a statistical procedure called *correlation*. Similar to its usage in everyday language, the word *correlation* indicates an association between two elements. Even if the statistics "work" by meeting preset levels of mathematical significance, the relationship is meaningful only if it makes explainable sense.

It's also important to consider another caveat to understanding statistical prediction and association: That is, we can never know that we're accounting for all the factors that come into play in a person's life. Even if we find, for example, that depression in childhood strongly predicts use of tobacco years later, it is only one of many variables in a person's life. In human behavioral research, we can never identify all the meaningful variables. Scientists can, at best, identify some that matter.

Another aspect of research that applies particularly to work with adolescents is the fact that usually studies are run and data are collected by adults. Unfortunately, what adults expect to affect a young person can be as unrelated to the adolescent experience as the Arizona T-shirt slogans. Another dud with adolescents was an ad in which a for-real "talking" camel announced its displeasure with being used as a symbol for smoking. Adults (including this author) found it clever. Focus groups of young persons, however, weren't amused.

Is the Outcome Inevitable?

Remember candy cigarettes? Remember how we used to "puff" on them as if they were real, all the time performing a grand, sweeping gesture with the hand that held the cigarette? Remember how adults used to tell us not to buy them, lest we become inured to the ways of tobacco and become smokers? Remember those olden days?

Actually, those days are right now. Candy cigarettes and others of their ilk still hold their own in the confectionary business, which also offers children cigars and pipes made from bubble gum, chocolate, and other candies. These sales are predictably controversial, enough so that at least one candy maker removed the "tip" from its candy cigarette and renamed it Candy Stix. Several states have tried, without success, to outlaw candy tobacco-look-alikes; some convenience stores refuse to sell them even if they are legal.

Just what draws a young person toward tobacco? If we had better answers to that question, perhaps we would have better ways to prevent pediatric tobacco use in the first place. Sometimes, it can help to reverse the question and determine what draws people *away from* tobacco. For example, the state of Utah, which is heavily populated by persons with a religious prohibition against smoking (Mormons), has the nation's lowest adult smoking rate at 13 percent. Even though Utah ranks low in youth cigarette use, some 44 percent of female and 52 percent of male Utah high school students have tried smoking. Those figures are lower than similar data from other states, but they still represent a substantial number of young persons. Fewer Utah youths reported current smoking when questioned for a 1995 survey (17 percent for both females and males), and less than half of those were considered frequent users. The teen smoking rates in neighboring Wyoming were more than double those of Utah, and in Nevada they were nearly double.

Does the anti-tobacco religious influence cause these regional differences? Is the youth smoking rate lower among children who are less exposed to adult use of tobacco? In contrast to the other western United States statistics, a greater percentage of young people in California reported having tried tobacco (63 percent of females and 65 percent of males), but only 7 percent of females and 8 percent of males were frequent cigarette users. Smokeless tobacco rates among young persons were also lower in California than in Utah. While California does have a sizable number of Mormons and other nonsmoking groups, such as Seventh Day Adventists, their representation among the overall state population is not sufficient to swing the numbers that dramatically. A more likely explanation is the aggressive anti-tobacco campaigns carried on in California in recent years, coupled with the basic California ardor for health.

A 1998 longitudinal report concluded that advertisements and promotions lead one-third of teenagers to try tobacco. John P. Pierce and his colleagues interviewed a sample of nonsmoking California adolescents in 1993 and re-interviewed them again in 1996. Although the teenagers stated in 1993 that they had no intention of smoking, those

Social Promotion

$1 billion	1983 U.S. tobacco advertising expenditure
$1 billion	1992 U.S. tobacco advertising expenditure
$1 billion	1983 U.S. tobacco promotional expenditure
$6 billion	1992 U.S. tobacco promotional expenditure
"sleek cat travel kit"	What Virginia Slims will send you free for mailing in 60 UPCs (barcodes)
"truly tribal bag"	ditto, for 80 UPCs
"leopard lingerie"	ditto, for 165 UPCs
"signature safari jacket"	ditto, for 315 UPCs
"go native pants"	etc.

who had a favorite cigarette advertisement in 1993 were twice as likely to later start smoking or be willing to start as were those with no favorite ad. Those owning or willing to use a tobacco promotional item in 1993 were nearly three times as likely to be smoking by 1996 as those who were unwilling to use a promotional item. About half of the nearly 1,600 teenagers sampled moved closer to becoming smokers between 1993 and 1996. Nearly 30 percent had experimented with smoking during the three-year interval. Of those who expressed a preference for a favorite ad in the first interview, 83 percent favored Camel or Marlboro.

Also, what pulls children and adolescents toward or away from smoking differs from group to group. Reports from Gilbert Botvin and colleagues at Cornell and Columbia universities, who studied predictors of smoking among inner-city Latino and African-American youth, indicated that the most important social influences promoting smoking were friends and peers. Additionally, feelings of hopelessness, lack of efficacy in basic life skills, and low self-esteem appeared to contribute to the likelihood of smoking. In this research, measures of socioeconomic status were unrelated to the extent of smoking. These reports are particularly pertinent because smoking rates among African-American youth are lower than the rates for whites until the trend flip-flops as both groups reach adulthood, when whites' smoking rates fall below those of African-Americans.

The concept of "group self-identification" has also been used to predict adolescent cigarette smoking. A longitudinal study by Steve Sussman and colleagues at the University of Southern California and the University of Illinois at Chicago noted that the peer groups with which seventh graders identified themselves predicted (statistically) smoking in eighth grade. These were the group categories as derived from previous self-descriptions by youth: (1) high-risk youth, including stoners, heavy metalers, and bad kids; (2) skaters, including skaters and surfers; (3) hotshots, including brains and socials; (4) jocks, composed of jocks and cheerleaders; (5) regulars, including new wavers and actors; and (6) others. The highest rate of smoking was among the high-risk youth. Although group self-identification and seventh-grade smoking did significantly predict smoking a year later, the authors were careful to clarify that self-identification was "a fair predictor" with its own merits, but it did not describe the total picture.

The predictive relationship between smoking and the negative emotions of depression or anxiety has been the object of considerable research in adults, but fewer studies have examined it in children and youth. Paul Rohde and a research team at the Oregon Research Institute determined that adolescents who had experienced an episode of depression experienced "psychosocial scars" that included cigarette smoking, in addition to increased health problems and excessive emotional reliance on others. "One implication is that although the rate of smoking in the general population may be decreasing, it may be increasing in adolescents who have experienced an episode of depression," they wrote. "Another possibility...is that...depressed adolescents progress from experimentation to more serious levels of tobacco use."

Children's competence and their parents' behaviors were linked to early tobacco use in research reported by Christine Jackson at the University of North Carolina at Chapel Hill and her colleagues. They found that children who rated themselves as less competent and whose teachers also rated them as less competent were more likely to use tobacco at an early age. Children were also more likely to use tobacco at an early age if their parents were nonaccepting and if their parents were less skilled at setting rules and supervising behavior. Additionally, children of at least one smoking parent were twice as likely to smoke as were their peers whose parents did not smoke.

Children's personality styles at ages six and ten contributed to early use of cigarettes, according to research by Louise Mâsse at the University of Texas-Houston and Richard Tremblay at the University of Montreal. They found that "novelty-seeking" and "low harm avoidance" predicted the early use of tobacco and other substances. They were investigating dimensions of personality postulated by Robert Cloninger in the 1980s. *Novelty-seeking* is believed to be an inherited tendency toward "exploratory activity and exhilaration" prompted by novelty, or by things that appeal to the appetites. *Harm avoidance* refers to an inherited tendency to "react intensively to aversive stimuli," which controls the learning mechanisms that enable inhibition.

Scientists attempting to understand the onset of tobacco use also watch for trigger events and environmental risk factors, in addition to vulnerable personality styles. Young smokers themselves state that they smoke because they enjoy it. A survey of more than 10,000 British adolescents, questioned by J.R. Eisner and colleagues in the 1980s, found that they attributed their cigarette use to the experience of smoking itself, rather than to peer pressure. They said that they found smoking to be an enjoyable, calming act that helped them deal with stress. They were more inclined to reject notions that they were smoking because it was a grown-up thing to do, or because of how it made them look among their peers.

University of Reading, England, researchers David Warburton and colleagues in 1991 summarized such findings to date: "It may not be simple exposure to nicotine that results in adolescent smoking, but that smoking results from the situations in which the young people find themselves at this most stressful time of life." The scientists identified smoking as "a coping strategy" for both younger and older smokers.

Evidently, many young white women use smoking as a way to control body weight. Findings from Memphis State University's Diane Camp and her fellow investigators indicated that nearly 40 percent of the surveyed adolescent female smokers stated that they used smoking to depress appetite and control weight. About one-fourth of young male smokers reported the same. Amazingly, not even one African-American teenager who was questioned, either male or female, reported using smoking to control weight or appetite. The researchers concluded that the racial differences might be explained in view of young white women's greater vulnerability to pressure that they be slender. Young African-American women, on the other hand, expressed less discrepancy between their actual weight and their ideal weight than do white adolescent girls. The person most likely to use smoking as a means of weight control was a white female who chronically diets, the researchers determined. They raised the possibility of an intertwining of depression, restrained eating, and smoking in young women. (See chapter 6.)

An Unnatural History

A natural history museum often endeavors to explain the course of events in the life of the Earth, such as the evolution of a dinosaur or a horse, tracing the species' change across time as it moves toward maturation or extinction. Tobacco use also has its own history. This history usually starts in childhood or adolescence and often progresses across the life span of a tobacco user.

A longitudinal study that followed more than 4,000 adolescents into adulthood found that smoking in adolescence was a powerful predictor of smoking in adulthood. Once the adolescents had become adults, many who were smokers attempted cessation and experienced relapse. Few individuals studied actually started smoking as adults. Researchers Laurie Chassin and colleagues from Arizona State University and Indiana University found that young persons with less education and with parents who smoked were less likely to quit smoking themselves. Those who adopted the adult social roles associated with marriage, parenting, and employment tended to quit smoking more than those who did not take on those roles.

The natural history of tobacco use is part of a larger picture of legal and illegal drug use. Columbia University investigators Kevin Chen and Denise B. Kandel followed 1,160 teenagers for 19 years, contacting them first in 1971 and then at three subsequent follow-ups. They reported that the major risk period for commencing the use of alcohol, cigarettes, and marijuana was before age 20, with cigarette initiation peaking at age 16. Smoking peaked in the early 20s and stabilized by age 22. The proportion of participants using tobacco had decreased only slightly by the time the participants were in their mid-30s. While heavy drinking decreased substantially in the study group as they aged from their 20s to their 30s, smoking showed virtually no decline.

Quoting Chen and Kandel: "Two behavioral features of drug histories continue to be strongly associated with persistence of use throughout adulthood: recency and frequency of use at an earlier period." Frequent use of tobacco at an earlier age predicted persistent use well into adulthood. Of all the substances measured, including so-called hard drugs as well as marijuana and alcohol, persistent use of cigarettes was one of the most serious drug-related health problems the study group faced.

The biggest risk of adolescent smoking is that it won't stop once adolescence ends. John P. Pierce and Elizabeth Gilpin of the University of California at San Diego posed the following question in the title to a report on the life span of smoking: "How Long Will Today's New Adolescent Smoker Be Addicted to Cigarettes?" Their answer, summarized, was this: "[T]hese data predict that smoking will be a long-term addiction for many adolescents who start now." Their findings can be coupled with those of Naomi Breslau and Edward L. Peterson of the Henry Ford Health Sciences Center, who determined that the likelihood of eventual cessation was higher in those smokers who started smoking after the age of 13. They suggested that efforts aimed toward delaying the onset of smoking could increase the potential for quitting, and thus could help reduce the death and disease associated with tobacco use.

Pierce and Gilpin found the escalating social undesirability of smoking to be encouraging, even though smoking has become increasingly acceptable among the young. Prevention programs could help teens who are still in the "initiation process" of becoming acquainted with tobacco and who have not yet become dependent smokers, the authors noted. They also placed a high priority on efforts to prevent initial experimentation.

Tobacco use in childhood and adolescence may predispose people to use other substances. Teenagers who drink alcohol and smoke cigarettes even infrequently are 30 times more likely to use marijuana than are those who neither drink nor smoke. Also, among teenagers with no other problematic behaviors, using tobacco, alcohol, and marijuana even to a limited extent increases the risk of other drug use (heroin, cocaine, LSD, etc.) by 17 times.

Cigarette smoking among U.S. adolescents was associated with risky sexual behaviors, marijuana use, binge drinking, and engaging in physical fights, according to a 1997 CDC analysis. Among young persons in Australia, tobacco and alcohol were both deemed to be "gateway" substances leading to other drug use, as reported by Debra Blaze-Temple and Sing Kai Lo of the National Centre for Research into the Prevention of Drug Abuse at Curtin University of Technology in Western Australia.

All of this may seem distant and irrelevant to someone who considers anyone over 21 to be ancient, and who suspects that turning 40 is a near-death experience. This perceived irrelevance is a major reason that smoking prevention programs often are ineffective with young persons and that cessation programs are difficult to design. Donald J. Reid and colleagues of the Association for Public Health in London summarized teenage smoking in Western countries: "Interventions aimed primarily at youth are likely to have a delaying effect only, and sophisticated school programmes, though potentially valuable, have proved difficult to implement effectively on a large scale." They proposed, instead, community interventions that covered all age groups, changes in fiscal policy, smoking restrictions, advertising bans, and media campaigns.

An article in the American Psychological Association's monthly *Monitor* about the ineffectiveness of substance-use prevention programs indicated that U.S. schools lack access to effective programs and don't use "the best prevention science." Those programs with good track records, including a Life Skills Training program developed at Cornell University Medical College and the Midwestern Prevention Project designed at the University of Southern California, are not widely used. They haven't been "touted to the public," as the author Bridget Murray lamented, laying responsibility on school administrators, researchers, and funding agencies. Shekeh Kaftarian of the U.S. Center for Substance Abuse Prevention's National Center for Advancement of Prevention stated, "You can't just pull any program off the shelf and hope it works."

Both people and tobacco, it seems, are more complicated than that.

The Spitting Image

While the prevalence of smoking—the most common nicotine-delivery system—has stabilized or decreased in many demographic groups in recent years, the use of smokeless tobacco has increased dramatically, particularly among teenagers and young adults of both sexes. The expanding popularity of smokeless tobacco products dates to about 1980, mostly reflecting increased consumption of moist oral snuff.

While the use of smokeless tobacco products has risen, public knowledge about smokeless tobacco has been marked by incorrect notions that smokeless tobacco use is an outdated practice, or a practice with negligible health risk. Compared to what they know about smoking, scientists have learned far less about smokeless tobacco. We know little about how its effects differ from those of smoked tobacco, for instance. We know it has the potential for dependence, as does smoked tobacco, but researchers have only begun studying the ways that dependence on smokeless tobacco differs from dependence on cigarettes. We do not yet know what factors come into play in fostering the continued use of smokeless tobacco over many years.

We probably know more about smokeless tobacco's history than we know about its effects. Tobacco has been used in many forms, including oral and nasal forms, in many cultures, from Sweden to southeast Asia, for centuries. Smokeless tobacco, which was both chewed and used as a nasal snuff, was common among upper-class Europeans. As the practice of spitting out saliva laden with tobacco juice came into disfavor because of its potential for spreading disease, smokeless tobacco use declined. Cigarette use, on the other hand, increased. Although the European Union in 1992 prohibited the sale of moist oral snuff in all countries except Sweden (where, according to Swedish scientist Gunilla Bolinder, it was regarded as a tradition), the prohibition was not yet implemented at this writing. The seesaw trade-off in tobacco products has continued; as cigarette usage has declined for the last several decades, smokeless tobacco use has increased.

If smoking is baffling to non-smokers, the use of smokeless tobacco is an even greater puzzle. Non-users of smokeless tobacco often have no idea how snuff is used, or why. One researcher of smokeless tobacco reported that such questions as these were common, even among smokers:

- Do people stuff snuff up their noses? (*Answer:* They can, but that isn't a common practice in the United States.)
- How do people use it? (*Answer:* Users park a "dip" or "chew" or "plug" in their mouths, usually in their cheek or alongside their gums, and the nicotine is absorbed through the lining of the mouth.)
- Does snuff consist of bags of loose tobacco leaves? (*Answer:* Usually it is sold in little cans or pouches. Those cans of Kodiak, Copenhagen, and Skoal are all smokeless tobacco, or "moist oral snuff," which is the most popular form of smokeless tobacco used

in the United States. Chewing tobacco, on the other hand, is sold in bags.)

- Isn't using snuff just something old people do, sitting on their porches somewhere in the Ozarks? (*Answer:* Not any longer. Some areas report use among teenage boys to be as high as 25 percent, and among girls about one-fifth of that or less. In some locations, such as some Native American reservations, as many as 70 percent of both men and women use smokeless tobacco.)
- Is it purely tobacco? (*Answer:* No, moist oral snuff consists of cut or ground tobacco leaves, water, and flavoring.)
- Do users swallow it? (*Answer:* Not generally; they spit out the tobacco-laden saliva.)
- Where do they spit? (*Answer:* Usually in a cup or other receptacle, but look around before you go barefoot . . .)

The majority of smokeless tobacco research has examined initiation of smokeless tobacco use in teenagers younger than 18 years old. Little research has focused on continued adult use. Consequently, we do not know whether the factors that led to initiation of the practice are involved in maintaining the behavior. We also do not know the role of nicotine dependence in maintaining smokeless tobacco use. Additionally, it is unclear to what extent research findings about psychological factors leading to smoking are applicable to smokeless tobacco use.

Smokeless tobacco involves a distinct pattern of use, often being consumed gradually over a period of hours. Nicotine's effects differ according to the speed and intensity of its absorption. This is believed to account for the differences in psychological impact between cigarettes and smokeless tobacco. Such factors as depression, anxiety, risk-taking, peer and family influences, and perceived health risk have been associated with adolescents' use of smokeless tobacco. University of Minnesota scientist Dorothy Hatsukami, an authority on this and other forms of tobacco, and her associates noted that precursors to smokeless tobacco use included depression and relaxation, and that the self-reported feelings of depression were directly related to feeling bored.

Other researchers have found that among smokeless tobacco users of middle-school age, anger is commonly reported.

Work by Elbert Glover and his colleagues indicated that college students using smokeless tobacco were likely to have family members who also used snuff. Users said that smokeless tobacco helped them relax. They considered it less harmful than cigarettes. The tendency to take risks and to get into trouble have been associated with smokeless tobacco use among adolescents in several reports, as has a lack of concern about negative social consequences of smokeless tobacco use.

None of the psychological factors that are believed to lead to the initial use of smokeless tobacco or to contribute to its continuing use are diagnosable psychological conditions. They are all merely normal human traits that vary in degree from person to person. In continued use, alleviation of feelings of depression may contribute to persistent use of snuff, but users of smokeless tobacco are not necessarily pathologically depressed. As is the case in many conditions that occur in human existence, the combination of otherwise benign vulnerabilities and situations may be what leads a person to years of using a potentially hazardous substance.

Some researchers and clinicians encourage smokers to shift from cigarettes to smokeless tobacco as a means of reducing health risks. Although the risks associated with smokeless tobacco use appear to be less than those associated with smoking cigarettes, they remain substantial. As Gunilla Bolinder concluded: "It seems beyond doubt that the use of chewing tobacco together with different mixtures of flavoring, alkaline substances, and a variety of natural products, when used in populations with poor mouth hygiene, and inadequate nutritional status, is causally related to the development of oral cancer."

And lest those not in developing countries think they are immune, consider data indicating that the relative risk of oral cancer is significantly higher in users of moist oral snuff. As for whether snuff use increases the risk of other cancers, the evidence is somewhat inconclusive. Nonetheless, Bolinder pointed out, even if smokeless tobacco is less risky than smoked tobacco, it presents considerably greater poten-

Chew 'bacca

(All statistics are U.S.)

4.8 million   adult men who use smokeless tobacco

533,000   adult women who use smokeless tobacco

up to 80%   Native American adolescent males using smokeless tobacco

up to 70%   Native American adolescent females using smokeless tobacco

20% - 25%   smokeless tobacco users who also smoke cigarettes

79%   smokeless tobacco users who started using by ninth grade

$100 million   amount by which sales revenues of smokeless tobacco grew annually in 1990s

25%   white high school males reporting using smokeless tobacco at least monthly

14 million   increase in pounds of smokeless tobacco sold anually from 1980s to 1990s

more than 50%   rate of smokeless tobacco use among varsity male athletes at two southwestern universities

4%   smoking rate among the same athletes

tial risk to health than such hazards as pollution, UV radiation, or tainted food. And, as he soberly added: "One must still keep in mind that the use of smokeless tobacco involves the exposure to one of the most addictive substances known."

Epilogue

And what of Douglas Spaulding?

At the end of Bradbury's 1957 book *Dandelion Wine*, Douglas is still invincible and all-powerful. As the deep of night descends, he commands the town to get ready for bed, to brush their teeth, and to turn out their lights. With his sleep, the summer ends. He is, after all, the mighty age of 12.

If only he had thought, during his moments of power, to suggest that all the smokers lay down their cigarettes and stop smoking.

CHAPTER 3

"The interesting realization that I went through last time I tried to quit was that I do not know myself as an adult without the drug nicotine," she muses. "I started smoking when I was 13 or 14, and I've been smoking for over half my life.

"My entire adult life has been under the influence of this drug. And I don't know myself as an adult without it." What would she be like, I wonder. Would her wit have a dull edge? Would her eyes dart a little more slowly, her gestures be a little more subdued? What would be different, or would I even notice the difference?

"I find that when I quit, I need to really work on behavioral changes as well, and not just for the withdrawal period, but also after the withdrawal has ended. Without smoking, I think differently, I feel different, I feel less patient. I get deeply irritated.

"I know I'm gonna have to go through and do it all again when I quit. I'll have a better starting place next time because I'll know more what fits and what feels good and what I feel comfortable with, and how to be patient, and how to keep my mouth shut (which I have to work on).

"I'm going to have to do it again. And that's scary."

Enhancer and Necromancer

Most of us use something, sometime or another. We want that little kick in the morning, that little oomph in the afternoon, that little flight away from blah, or that little sharpening on a dull day. To get that little buzz, we drink that cola, sip that coffee, nibble on that chocolate, or smoke that cigarette. Unless we ingest caffeine in fairly high doses, we don't get as much of a jump-start from it as we would get from nicotine. The effects from the nicotine in smoked tobacco take about seven seconds to occur. Achieving the same effects from caffeine takes 10 to 20 minutes. Some researchers believe that smoked tobacco's almost immediate impact is what makes its effects so reinforcing.

It is nicotine's adaptability, as well as its addictive properties, that make it so popular. A smoker not only forestalls abstinence symptoms, or withdrawal, with the morning drag, the smoker probably also perceives that he or she is affected emotionally and mentally. This is accomplished not through using one brand with a precise level of nicotine in its cigarettes, but through a smoker's precise ways of using tobacco to achieve a desired state.

Whether the nicotine in tobacco enhances mental skills is a focus of scientific debate. On one side is the argument that nicotine enhances specific aspects of thinking skills, or cognition. On the other side is the hypothesis that the so-called enhancements are really just "medicating the deficit" that occurs when a smoker is in withdrawal. As in many efforts to study human behavior and performance, this debate hinges to some extent on questions of methodology. And as with many explorations into human behavior, it starts in animal laboratories.

Researcher Karey Elrod and colleagues at the Medical College of Georgia reported a decade ago that nicotine enhanced task performance in young adult macaques (nonhuman primates). The macaques' performance on a matching task improved when they were administered nicotine. Conversely, chemically blocking the nicotinic receptors in the animals' central nervous system inhibited performance. The researchers hoped to extend their research to humans suffering from dementia.

Nonhuman research has covered such smoking-related factors as the effects of prenatal nicotine exposure on rat pups (i.e., baby rats), a line of inquiry that allows researchers to explore the subtleties of prenatal nicotine exposure. The adverse effects of a mother's smoking during pregnancy are a health concern because about one-fourth of pregnant women in U.S. urban areas smoke. Their tobacco use affects their offspring's size, health, and behavior, often throughout the child's life. Studies also have noted long-lasting cognitive deficits that appear to result from prenatal smoking.

An animal study that examined analogous effects in rat pups was performed by Edward D. Levin and colleagues at the Duke University Medical Center. They exposed pregnant rats to nicotine infused through a tiny implanted pump, and later monitored the offspring's performance on maze tasks. The exposed rats' performance was compared with that of control rats that received no nicotine exposure. The investigators found subtle but measurable effects of prenatal nicotine exposure. The nicotine dosage was low enough not to cause any significant deficits in the mother rats' weight gain, in the litter size, or in the pups' birth weight. However, it was sufficient to affect performance.

The researchers were careful to note that although the infusion method was useful for the study question, it did not exactly parallel the nicotine effects a smoker would experience over the course of a typical day of smoking. Despite that limitation, they concluded that their studies provided evidence that prenatal nicotine exposure could increase the vulnerability of offspring to lasting deficits, and could decrease any benefits from being in an enriched environment after weaning. They also expressed concern that prenatal nicotine exposure could reduce the adaptability and development of an offspring's nervous system, which could impair chances for recovery from damage caused by the nicotine exposure.

Enhancer

The next step, following the lines of animal research, has been to extend the findings to humans. One of the notable efforts to analyze the effects of nicotine on human cognition and task performance was a 1993 report from Stephen J. Heishman and colleagues of the National Institute on Drug Abuse's Addiction Research Center (NIDA/ARC) in Baltimore. Rather than follow a more typical course of testing smokers who had been using tobacco for some length of time, Heishman's group used volunteers who were naïve to tobacco and thus had little or no experience using nicotine. The hazard in the study was that exposing the participants to nicotine might put them at risk for later taking up regular tobacco use. This risk was mitigated to some extent by the subjects' using exact amounts of nonsmoked nicotine, starting with a placebo (non-nicotine) dose, then a 2-mg dose 90 minutes later, then a 4-mg dose 90 minutes after that. The researchers measured physiological measures including heart rate, blood pressure, and skin temperature, as well as assessing performance on several cognitive tests. In this group of nonsmokers, nicotine did not enhance test performance.

Heishman reported in 1996 the results of a second study, in which

Well Blended

2.3%	starting percent of nicotine in tobacco blend for U.S. Patent No. 5,065,775
5.2%	final percent of nicotine in tobacco blend for U.S. Patent No. 5,065,775, following processing
2.6%	starting percent of nicotine in flue-cured tobacco, U.S. Patent No. 4,898,188
4.8%	final percent of nicotine in tobacco blend, U.S. Patent No. 4,898,188, following processing

(Kessler, 1994.)

he examined nicotine's cognitive enhancement effects on 12 tobacco-naïve men who stayed in an inpatient research unit and were given various doses of nicotine gum over nine days. The men's performance was actually impaired by nicotine, particularly at the highest dose level. Although nicotine enhanced response time on some testing, it also impaired accuracy. Again, the data did not support the hypothesis that nicotine enhances cognition in nonsmokers.

Jacques Le Houezec at Hôpital de la Salpêtrière in Paris and collaborators at the University of California at San Francisco also studied the effects of nicotine administration in nonsmokers. The team evaluated the performance of 12 nicotine-naïve research subjects before and twice after injections of nicotine or injections of a saline solution, along with a nontreatment (or control) condition of no injections. As would be necessary in such research, they balanced the order of the substances (nicotine given first to some subjects, saline given first to others) to control for any possible effects that the order of administration might have, as well as to help control for practice effects in repeated testing. The injections were given in a "double-blind" condition, in which neither the subjects nor the person administering the testing or the injection knew which substance was in the injection. The researchers also administered an EEG-like procedure called an *event-related potential*, in which brainwave measures were taken repeatedly to assess the subjects' responses.

Le Houezec and his colleagues found that testing given 15 minutes after the administration of nicotine showed significant evidence of cognitive enhancement, but that the enhancement was no longer significant 45 minutes after the injection. They also noted that nicotine increased the number of responses but did not increase accuracy.

Why study nonsmoking, nicotine-naïve subjects? As tolerance develops to a substance such as the nicotine in tobacco, the effects may differ at varying dose levels. By definition, tolerance involves the need for increasing doses of the substance to achieve the desired effect. If nicotine is a cognitive enhancer, and if for the sake of consistency all research subjects must be given the same dose (relative to body weight), then a low dose might not produce the same effect in a dependent smoker that it would yield in a nicotine-naïve subject. Even among a group of dependent smokers, it is unlikely that the same dose will have the same effect. Therefore, it becomes important to use nicotine-naïve participants for some experiments.

A drawback to using nicotine-naïve participants is that nicotine's fundamental toxicity, which makes a teenager nauseous with that first cigarette he or she tries, has the same effect on adult participants in research studies. A limitation to such research, as it turns out, is not that naïve research volunteers will take up smoking, but that they will have difficulty using nicotine without feeling too ill to participate. Fortunately for the sake of research paradigms, nicotine doesn't affect every first-time user quite the same way.

An additional methodological problem comes in comparing data from smokers with data from those who have never smoked or who are former smokers. Never-smokers and ex-smokers are not necessarily equivalent to smokers, and thus comparisons must be made cautiously. David G. Gilbert of Southern Illinois University has recommended randomly assigning smokers to intermediate-term abstinence from tobacco, and offering them a financial incentive sizable enough to compensate them for their trouble and to minimize attrition from the research study. This recruitment strategy would enhance researchers' capacity for making inferences about the effects of nicotine on performance independent of withdrawal symptoms. Gilbert also has recommended that researchers study mood in conjunction with cognition and performance, since mood can affect both.

An important question to address is whether nicotine enhances performance and cognition when it is administered after withdrawal. Frederick R. Snyder and colleagues of NIDA/ARC determined that when smokers were deprived of nicotine for 12 hours, their performance on a computerized battery of tasks worsened. Performance improved when they used nicotine gum containing doses of either 2 or 4 mg; in fact, it returned to baseline predeprivation level. The researchers concluded that the performance drop was specific to nicotine deprivation when the subjects abstained from tobacco longer than would be normal for them. The nicotine they received in the experiment treated their abstinence symptoms and served as a replacement for the nicotine in cigarettes, thus accounting for their improved performance when they used nicotine gum.

Methodological issues also come into play in various other studies that have attempted to look at nicotine's cognition-enhancing properties. A common criticism among scientists (a criticism they level at their own work as well as at colleagues' work) is that typical cognitive and performance testing bears little relationship to performance demands in the outside world. As George J. Spilich of Washington College argued, "Much of the research which investigates the effects of nicotine upon cognitive performance has been conducted with a very restrictive set of tasks." It is important, Spilich insisted, that addiction researchers choose measures that utilize the same processing demands encountered in the everyday world.

Tar, Baby

10 mg	amount of tar in a Marlboro Light cigarette, based on machine tests
18 – 36 mg	amount of tar a smoker actually could get from a Marlboro Light
0.7 mg	amount of nicotine listed for a Camel Light, based on machine tests
1.3 to 2 mg	amount of nicotine a smoker actually could get from a Camel Light
13%	smokers of ultra-light cigarettes who know the tar level of their preferred brand
1%	smokers of regular (full-flavor) cigarettes who know the tar level of their preferred brand
two-thirds	smokers who think that this information is listed on cigarette packs
3% – 8%	smokers who know that this information is listed in advertisements instead of on cigarette packs
www.cancer.org	World Wide Web site where the state of Massachusetts posts nicotine levels
nowhere	where tar and nicotine information is published for nonadvertised, generic brands of cigarettes

(See Kozlowski et al., 1998)

That should be simple, right? Why not just have research subjects attempt to program a VCR, insert wood screws in an assemble-it-yourself table, or navigate into a tight parking spot? Part of the problem is that tasks must be uniform, and performance must be measurable and repeatable. It is true that programming a VCR could be a timed task, but it is a task that some research subjects would already be adept at and that others would have an aversion to. Any prior experience programming a VCR could either enhance task performance or hinder it, or perhaps could add to the feelings of aversion. In any case, prior experience would make the findings difficult to interpret accurately. This is why it is important to use tasks that the subject might not have performed before, but that the subject can readily learn. That said, Spilich's point is well taken: For the research findings to be more meaningful, the tasks need to have some generalizability to tasks in everyday life. Determining that someone using nicotine can perform a tapping gesture faster than someone without nicotine might have little direct applicability to life's typical demands.

Neil Sherwood of the University of Surrey in England addressed such concerns by studying the effects of cigarette smoking on a one-hour computer-based driving simulation that involved tests of continuous tracking and brake reaction time tests. Sherwood investigated the effects of nicotine on smokers' driving-simulation performance while varying the nicotine levels in the cigarettes offered to study participants. He found that brake reaction time improved at all nicotine dosage levels, and that the consumption of two moderate-nicotine-level cigarettes enhanced tracking accuracy. He stated that, in retrospect, it would have been useful for him also to have studied the pre-testing baseline usage of tobacco in the subjects, and to use that information statistically. Nonetheless, he concluded that nicotine did have some performance-enhancing effects. He did not, however, advocate tobacco use as a means of sharpening one's wits. Rather, he concluded that smokers use nicotine as "an aid to concentration and to offset the negative effects of fatigue or boredom in everyday tasks like car driving." Using nicotine for these purposes could have implications for many activities requiring vigilance or attentiveness, conducted at home and at work every day—and could also indicate possible hazards associated with performing these activities during the early, difficult days of abstinence.

Sherwood's research represented an effort to explain demographic data indicating that smokers are overrepresented in motor vehicle accidents. Sherwood suggested that the "most cogent explanation" for the higher accident rate was the propensity of smokers to engage in risk-taking behaviors, coupled with smokers' tendency to use alcohol.

In 1992, Sherwood examined performance in smokers after single and repeated doses of nicotine gum. He found that smokers' performance on several tasks improved with the administration of nicotine. Among the measures showing improvement were reaction time, tracking, and memory reaction time. He interpreted the findings as suggesting that the enhanced motor performance smokers experience after the first cigarette can be maintained by repeated smoking. The goal of the study, however, was not to encourage repeated smoking of cigarettes, but rather to profile nicotine's *psychoactive* (psychologically active) nature.

Researchers have been particularly interested in examining the effects of nicotine on attention and memory. Although at first glance these two aspects of human thought processes may seem to differ, they are sufficiently connected to require careful examination and separation. A person is unlikely to remember something to which he or she had paid little or no attention. In fact, what appear to be memory deficits among people with reversible and treatable conditions such as depression might instead merely reflect their difficulty in focusing attention.

An example: A (nonsmoking) woman who was an enthusiastic and avid reader experienced a depression lasting several months during a period of unexpected unemployment. She read many books during this period, passing the hours and days between job interviews and social contacts. Among the books she read was a science fiction work by Ursula LeGuin, an engaging and talented writer. Several years later, long after the woman was reemployed and her mood had dramatically improved, she saw a televised dramatization of a LeGuin novel. She wondered how, in her exhaustive reading of LeGuin's works, she had missed that particular story. She found the book tucked away on a bookshelf in her home and noted with surprise that she had not only read the book several years earlier during the job-hunting months, but she had underlined in it and made marginal comments. Her depression had so thoroughly unhinged her attention and focus that she not only did not remember what was in the book, but she didn't even recall ever reading it.

Gilbert, in his 1995 book *Smoking*, identified aspects of nicotine's cognitive enhancement that could play a role in a smoker's everyday life and could be affected by quitting tobacco use. Among the effects he categorized were these:

- *Visual Thresholds:* In smokers who are temporarily abstinent, nicotine abstinence lengthens the time at which a rapidly repeated stimulus such as a light flicker or light flash can be distinguished from the next flicker or flash. This distinguishing speed also is diminished in Alzheimer's patients.

- *Reaction Time, Speed, and Accuracy:* Abstinent and nonabstinent smokers who are given a dose of nicotine experience a decrease in reaction time in tasks requiring rapid responding. Performance on memory tasks, calculations, and other tasks generally increases in smokers who are likewise given a dose of nicotine. Recall of information does not appear to be facilitated as consistently by the use of nicotine. Nicotine administration in a laboratory setting has been shown to enhance memory functions in nonsmokers.

- *Usage of Different Parts of the Brain:* Gilbert and others have proposed that nicotine tends to cause arousal in the left part of the brain, which is associated with verbal tasks. This *lateralized*, or left-vs.-right, effect might also explain nicotine's impact on feelings. (See chapter 5.) The tasks that nicotine appears to enhance in smokers are believed to be mediated by the left side of the brain; conversely, functions mediated by the right side of the brain, particularly negative mood states and visual-spatial processing, might be inhibited by nicotine.

Check the Ingredients

more than 4,000	number of identified chemical compounds in tobacco smoke
more than 50	number of identified cancer-causing chemical compounds in tobacco smoke
carbon monoxide	biological marker of recent smoking; measured in exhaled breath
carboxyhemoglobin	biological marker of tobacco use, measured in blood
cotinine	product of nicotine metabolism, measured in saliva, blood, or urine; half-life of 16-20 hours allows measurement of recent nicotine use
thiocyanate	biochemical marker of tobacco use; long half-life of 14 days allows measurement of less recent nicotine use
pretty low	percentage of smokers worldwide who have heard of carboxyhemoglobin, cotinine, and thiocyanate

Nicotine's usefulness as a cognitive enhancer is limited. Researchers have found that nicotine use can result in what is called *state-dependent learning*. In other words, what is learned under the influence of nicotine can be optimally utilized *only* under the influence of nicotine. This means that a student who smokes all night while he crams for an exam the next morning might find himself unable to remember the material unless he has a comparable amount of nicotine in his system when he takes the exam. Since smokers generally attempt to maintain somewhat steady blood levels of nicotine throughout the day, this might not be a problem under most circumstances. A three-hour exam with no smoking break, on the other hand, could present a problem.

David M. Warburton and colleagues of the University of Reading specifically examined this issue in 1986 in relation to nicotine use. They designed two studies, one with cigarettes and one with nicotine tablets. In the first study, they tested recognition of a visual stimulus by smokers who had been primed by smoking before engaging in the learning task. As a contrast to prior research utilizing word recall, the investigators used Chinese ideograms that would be difficult for subjects to translate into any form of language (the subjects apparently did not know the Chinese language). Subjects were tested in a variety of conditions in which they were told either to smoke or not to smoke before learning, during learning, and during recall. The researchers found that nicotine facilitated the input of information and that the learning was indeed state dependent.

In a second study, Warburton and his associates examined the effects of nicotine on registration and recall of verbal material. Smokers who came for testing after overnight nicotine deprivation were instructed to learn and recognize 48 words from 12 categories such as fruits, animals, or musical instruments. Participants were given nicotine in tablet form with tabasco sauce added to mask the flavor, so that the subjects wouldn't know whether they were getting a nicotine tablet or a non-nicotine placebo tablet. Again, the researchers found that nicotine facilitated learning and produced state-dependent learning.

Such features of nicotine use are part of what keeps smokers smoking. These features are also part of what makes quitting so difficult. Regarding quitting, researchers have become reluctant to write and speak directly of nicotine *withdrawal* per se, preferring instead to use the term *abstinence effects*. Their caution comes from a concern that not everything a smoker experiences when going without the usual amount of tobacco is necessarily related to withdrawal from nicotine dependence. Rather, the term *abstinence effects* is a larger umbrella that can include both the symptoms of physical withdrawal and the impact of being deprived of the substance's effects. An analog to this condition could be the situation of a wife who loses her husband. Not only is she in mourning because of the loss of her husband, but suddenly now she does not have anyone to help carry in the groceries and carry out the trash. The loss itself causes a grief response that is separate from having to do those tasks by herself. To compare it with smoking, the withdrawal per se would be the grief response; the sudden lack of help with the groceries and garbage would be analogous to the abrupt absence of the effects of the substance.

Both types of effects are pronounced when a tobacco user is abstinent. From within the experience, it is probably difficult (and pointless) to try to sort out which noxious effect is due to which specific cause. For multiple reasons, a smoker's cognition and performance tend to become impaired during abstinence. The effects can last as long as several weeks, although most smokers' abstinence effects subside within two weeks. Other conditions, such as the smoker's overall health, stresses, and even (for women) menstrual state, can interact with abstinence effects to influence cognitive capacity.

In a nasty twist suitable for a Dostoevsky novel, researchers Todd M. Gross and colleagues at the University of California at Los Angeles and the West Los Angeles Veterans Affairs Medical Center found that abstinence not only makes the heart grow fonder, it results in a sort of fixation on smoking. Some scientists had suspected for years that using an addictive substance over a long period of time increases responsiveness to information about the substance. For example, alcoholics tend to notice drinking-related concepts faster than nonalcoholics do. This behavior becomes virtually automatic, requiring no conscious effort. When a substance-dependent person then becomes abstinent, the automatic behavior is "frustrated," and thoughts about the desired substance begin to intrude.

This phenomenon was demonstrated largely by anecdotal evidence until Gross and colleagues set out to investigate it in smokers. They used a neuropsychological test called the Stroop, in which a participant is asked to state the color of ink in which a certain word is printed. To successfully perform the Stroop task, the person must suppress the meaning of the colored word to state its color. Sometimes words intrude to the point that suppression is difficult, requiring concentration and effort. When a person's focus is fixed on a particular concept, this slows the naming of the colors of words that depict the concept. For instance, someone with an eating disorder is likely to name the colors of food-related words more slowly than he or she would name the colors of neutral words.

In the study conducted by Gross, abstinent smokers required more time to name the ink colors of smoking-related words than of neutral words. The investigators suggested that this meant that nicotine abstinence among smokers resulted in some type of preoccupation with smoking. This preoccupation "captures attention" and results in slower performance and interference with the task. They determined that this attentional shift was not due to any cognitive deficits associated with nicotine abstinence. These effects occurred only twelve hours into abstinence. Oddly, the abstinent smokers did not report having more thoughts of smoking than were reported by a complementary group of nonabstinent smokers who performed the same task. The difference showed up only in the results of the testing.

Help for Dementia?

A 1996 conference at Howard University brought together many of the world's experts in the cognitive and performance effects of nicotine. The group included researchers funded by governments, pharmaceutical companies, and tobacco companies. They debated the question of nicotine's enhancing qualities, not attempting to reach a consensus but rather to enhance one another's understanding. For two days, this group of scientists gave half-hour presentations on their latest lab findings. Some talked about the effects of nicotine on brainwaves, others about the effects of nicotine on mood and, indirectly, on cognition. Some explained the limitations of the various methodologies. Some clearly

Facts in Black and White

27%   U.S. whites (non-Hispanic, non-Native American) who smoke

35%   U.S. African-Americans who smoke

18 – 25 years   age range in which smoking is higher among whites than African-Americans

25 – 65 years   age range in which smoking is higher among African-Americans than whites

10%   African-American smokers who smoke more than 25 cigarettes/day

one-third   proportion of white American smokers who smoke more than 25 cigarettes/day

75%   percent of African-Americans who smoke mentholated cigarettes

23%   percent of white Americans who smoke mentholated cigarettes

1.4 times greater   increased risk of cancer death for African-Americans vs. whites

1.5 times greater   increased risk of heart attack death for African-Americans vs. whites

1.8 times greater   increased risk for stroke death for African-Americans vs. whites

had a point to make, although none appeared to have too much of a bone to pick.

To those and other scientists, nicotine is neither good nor bad; it simply *is*. Human use of it may have good or bad consequences, but the substance itself and its effects are topics of inquiry, not matters of moral crusade. In that framework, many researchers have been encouraged and even excited to learn that nicotine may have properties that help patients afflicted with Alzheimer's disease. Early in the 1990s, reports began to be published to the effect that Alzheimer's symptoms might be treatable, to a limited extent, with nicotine. Research both at Duke University and at London's Institute of Psychiatry showed that some aspects of the cognitive deficits associated with Alzheimer's could be attenuated—not eliminated, not cured, but perhaps helped to some degree—by the administration of nicotine.

British researchers G.M.M. Jones and colleagues reported in 1992 that although Alzheimer's patients' short-term memory was not enhanced by nicotine, the perceptual and visual attention deficits that are part of the disease symptoms were improved by the subcutaneous (under-the-skin) administration of nicotine. They compared a group of young normal adults and a group of older normal control subjects with a group of 22 patients diagnosed with relatively early Alzheimer's disease. The tasks measured how well the subjects processed new information, attended to a task, engaged in fine-motor finger-tapping, and exercised short-term memory. The researchers noted that the Alzheimer's patients already had "large perceptual and attentional impairments" that accounted for substantial inability to perform the testing adequately. The subcutaneous administration of nicotine was useful in the research setting because it provided a safe way to administer nicotine in a study of the "acute," or immediate and short-term, effects of the substance. Nicotine from a transdermal patch would not have provided an adequate acute dosage for the research study, although the researchers indicated that it would be a preferred route of administration for a therapeutic dose. Also, it could be difficult to teach Alzheimer's patients to chew nicotine gum correctly, since the gum must be chewed for a short time and then "parked" in the mouth to allow the nicotine to be absorbed. This technique might be beyond the capacity of a dementia patient.

Additional evidence that nicotine might be therapeutically useful with dementia patients came in 1997 from Paul A. Newhouse and his collaborators at the University of Vermont and New York University. They explored the role of the central nervous system's nicotinic systems, or nerve receptors that are sensitive to nicotine, noting that one of the hallmarks of both Alzheimer's and Parkinson's diseases is a loss of nicotinic receptors in the central nervous system (or CNS, which consists of the brain and spinal cord). The team found age-related and disease-related alterations in several cognitive domains associated with CNS functioning. Their findings supported the belief that intact CNS nicotinic mechanisms are important for normal cognitive functioning, and that some of the cognitive deficits seen in Alzheimer's and perhaps Parkinson's patients could be due to the loss of CNS nicotinic receptors.

The Newhouse group used a substance called mecamylamine, which is a nicotinic *antagonist*, a substance that blocks the effects of nicotine on the CNS receptors. The results of administering mecamylamine, thus blocking the nicotinic receptors, mimicked the effects of a dementing illness. Their work suggested that such dementia-related problems as impaired capacity to acquire, process, and store information in memory could be related to the effects of a dementing disease on nicotinic receptors. Likewise, problems with attention, visual perception, and response speed could also result from inadequate CNS nicotinic reception.

This growing collection of research indicates that it might be possible to improve information acquisition and decrease cognitive errors in Alzheimer's patients through a process of nicotinic stimulation. The ongoing inquiry involves examining possible clinical benefits of nicotinic augmentation.

Scientists also have considered the possibility of using nicotine to treat non-Alzheimer's age-related memory impairment. Gary W. Arendash and associates at the University of South Florida examined the effects of nicotine injections on aged rats. (An aged rat, by the way, is 22 to 24 months old.) Older rats demonstrated severe learning and memory impairments, which were measured by the rats' ability to learn and perform tasks. The elderly rats were compared with younger rats in their ability to utilize a pole-jumping apparatus equipped with a grid floor capable of delivering a mild but aversive electrical shock. The rat had to learn to jump onto a pole within a five-second interval to avoid the shock. The rats also had to learn to run a maze that involved alternating left and right turns at each of six doors. (Do not leak this information to laboratory rats; they are supposed to learn it for themselves.) A third rat task involved determining which eight arms of a maze contained rat food, from among seventeen possible arms of the maze.

As the investigators pointed out in their published report, the aging process impairs learning and memory in rats and other animals. The researchers found that administering nicotine to rats in a regular pattern attenuated but did not eliminate the normal aging impairments of the rat brain. They were encouraged that nicotine demonstrated a broad ability to enhance cognition across diverse tasks that involved avoidance, cognitive "mapping," and spatial discrimination. They also noted that although nicotine had been administered to the rats for as long as a month, none of the rats evinced what is termed *behavioral desensitization*, or diminished effects of nicotine over time. In contrast, the aged rats were still performing well at the end of the study.

A 1995 study from Minnesota researchers A. Lynn Wilson and colleagues found that learning was enhanced in Alzheimer's patients using nicotine patch treatment. They exposed six patients with Alzheimer's disease to nicotine via a nicotine patch, also comparing their performance when they were using a placebo nonnicotine patch and when they were in a "washout" period in which the nicotine was clearing from their bodies. The researchers used a 22-mg nicotine patch, the highest dosage commercially available, which delivered about 0.9 mg nicotine per hour continuously over a 24-hour period. This dose level was considered intermediate.

The patients, who were mildly to moderately affected by the progression of Alzheimer's disease, were all nonsmokers. Their perfor-

mance task involved learning the placement of items within a multi-compartment box and following a sequence in locating the items. Investigators also monitored the patients' global cognitive functioning and their behavior in categories reflecting mobility, communication, inappropriate behavior, compliance, and aggression. The use of nicotine improved the patients' learning, as demonstrated on the experimental task, but did not alter their overall cognitive performance and behavior. The research team noted that the patients tolerated the nicotine administration well and experienced minimal side effects. Possible drawbacks to nicotine administration had been suggested in earlier research in which nicotine appeared to bring on increased anxiety and depression in some participants. Two of the Wilson study patients became belligerent and refused to participate in some testing sessions. Whether this was related to their use of nicotine was unknown. The patients also experienced some nighttime sleep disturbance. The investigators noted that any therapeutic use of nicotine should involve careful attention to achieving an optimal dose, which would vary from patient to patient.

No ethical researcher would suggest that patients take up smoking. But since nicotine is now widely available over-the-counter in safer forms than smoking, its therapeutic use has been considered seriously. In sum, now we know that nicotine can enhance some aspects of performance and cognition, at least in long-term use and in abstinent smokers. Nicotine might even offer some relief to dementia patients. This chapter should have a happy ending, right?

As encouraging as nicotine's potential therapeutic uses may be, its continued use by parents who smoke around their children has a deleterious effect on the children's thinking abilities. Even if the parents' cognitive skills might be maintained or in some ways temporarily enhanced through the administration of nicotine, their children do not fare so well. Exposure to environmental tobacco decreases children's performance on some cognitive performance testing, according to work by Karl E. Bauman and colleagues at the University of North Carolina. However, their research indicated that other factors in addition to environmental tobacco exposure also may have caused smokers' offspring to perform poorly on cognitive tests. Parents' mental capacity could be diminished by deleterious health effects from years of smoking and could thus make parents "less effective agents in the cognitive development of their children."

Necromancer

The power of nicotine to foster physical and emotional dependence may seem mystical or even magical. Perhaps nicotine's cognitive and performance enhancement potential is not very different than that of the morning cup of coffee or the mid-afternoon splash of cola. The enhancement itself, though measurable, is not monumental or even marked. The smoker may notice the difference, but others may not. Nicotine isn't a smart-pill that turns a mediocre thinker into a genius. It offers merely a temporary edge, and that edge might come only with repeated use. In any case, for the smoker it is an edge that also comes with a substantial risk.

The magic, it would appear, can be both black and white.

CHAPTER 4

S he stops to scrub her hands on the way back indoors after her mid-morning cigarette. As she washes, she explains that she doesn't want to smell like a smoker.

She puts her cigarette lighter back in her purse and we settle back into our chairs in her office. I ask how she knew it was time to smoke a few minutes before.

"I wanted a cigarette," she says. "I felt like it for about a half an hour before I told you, twenty minutes maybe. How do you know when you get thirsty? It's a part of your life if you're an addict, as I am. It's just a part of your life. You don't even think about it."

What does it mean to think of herself as an addict?

"I'm not proud of it at all. Many times, I wish I weren't an addict. Any addiction rules your life. It rules my life. If I can't have a cigarette, I start getting irritable, cranky, and unmanageable." She growls, but cutely. "It's like a bad hair day."

So, I ask, do you structure your day around smoking?

"I have one before I come to work, one once I get to work, one at 10 or 10:30. I eat lunch and I have one after I eat. I have one anywhere from 2:30 to 3:30, somewhere around in there.

"Here's the thing," she continues. "I go back to the fact that it's an addiction. Every person who is addicted has times when they're embarrassed or disgusted with their addiction. There were times when I was embarrassed to be a smoker, because of who I was around. I'd think, oh, God, if they find out I smoke, then they won't like me anymore, because smoking is disgusting. Which it is. They won't want to hang out with ashtrays. They won't like that smell.

"I get disgusted with myself because I find that so far I haven't been strong enough to beat the damn thing. But mostly I'm embarrassed because I feel like it's a weakness." She pauses, considering what she has just said. Then she reiterates, with emphasis: "It's a weakness."

And does she ever have the "I can't believe I ate the whole thing" feeling?

"I guess I've just kinda gotten used to that. I've smoked when I was out with friends at night, and then the next day I've said, 'I can't believe I smoked so much last night!' But it's never been that I was amazed at the amount; I was amazed that I was so unaware of it."

It is different, she knows, for some people.

"I have a couple of friends who are social smokers. They smoke mostly if they go out, they smoke once or twice a day or once or twice a week. I wish I could do that, and I can't. Once I quit, I can't ever let that happen again."

So that's something you learned from quitting? I ask. You learned that once you quit, you can't even let yourself smoke one or two cigarettes socially?

She shakes her head. "I tried that, a long time ago. It's like being an alcoholic. Stay away from that first drink. All the others don't matter. It's that first one."

Myths and Mysteries of Addiction

When a row of tobacco executives held up their right hands and testified to a U.S. Congressional committee that nicotine is not addictive, many people were displeased, but few were surprised. It wasn't new news. Dog bites man. Ho hum.

But when, in 1997, a tobacco company conceded that tobacco is addictive and that the company had known it all along, it was news indeed. Man bites dog, and the ensuing reports even made the cover of *Time* magazine. Never mind that millions of people, among them millions of smokers, already believed that nicotine is addictive. The news was that the tobacco industry was beginning to admit what many of the rest of us already knew. By early 1998, tobacco company executives stated that they knew it, too.

Oscar Delgado knew. An ex-smoker, he participated in the second 1996 presidential debate. He had one question to ask a candidate. As America watched, Delgado rose from the back of his section and addressed the candidate. "About 30 years ago, I was a pack-plus-a-day man, okay?" Delgado addressed his question to former Senator Bob Dole, who would later lose the election.

"You mentioned in a statement . . . some time ago that you didn't think nicotine was addictive," Delgado commented. "Would you care to . . . hold to that statement, or do you wish to recant, or explain yourself?"

"Oh, that's very easy," Dole replied, citing his voting record. Then returning to the question, Dole mused, "Are they addictive? Maybe they're—they probably are addictive. I don't know. I'm not a doctor." It was, he said, a "technical question."

His comment was pondered by many analysts, among them a psychiatrist writing for the *Wall Street Journal* the following summer. She provided the answer that, she claims, one pundit wished Dole had said: "Clumsy me," she parodied the response Dole should have given, "I got all tangled up in a technical matter—the nature of addiction—when all that I meant was . . ."

That wasn't the answer Dole gave, although it does consider the technical nature of the question of addiction. Behavioral scientists, psychopharmacologists, clinicians, and others studied this issue for decades, trying out this definition and that, before settling into what are now useful, if contested, terms.

The Elusive "Addictive Personality"

In Stephen Vincent Benet's epic *John Brown's Body*, the fictional belle Sally Dupré was like a forbidden fruit to young Clay Wingate, who fancied her. A soldier in the Civil War, Clay knew that if he so much as kissed her once before going off to fight, his heart would be unalterably turned toward her. "Your mouth is generous and bitter and sweet," Clay mused. "If I kissed your mouth, I would have to be yours forever." To a degree, this image depicts the fear of many people about addiction. They blame not the substance itself, but rather themselves for being addiction prone. They decide, with grim acceptance, that there must be something wrong with their personality.

Gotcha!

50%	younger adult U.S. smokers who are nicotine dependent (addicted)
87%	older adult U.S. smokers who are nicotine dependent
three-fourths	U.S. smokers who say they are addicted
78%	U.S. smokers who say they could quit if they decided to
two-thirds	U.S. smokers who say they wish they could quit
four-fifths	U.S. smokers who say they wish they had never started

A common cultural belief holds that some people are drawn like magnets to abusable substances and potentially compulsive behaviors. One drink, and they're alcoholic. One smoke, and they're addicted to nicotine. Whatever the substance, compulsion, vice, or temptation, it is supposedly the nature of their personality or their chemistry to be drawn toward it, to embrace it, and to be unable to abandon it. Believers in this notion apply the concept to an array of human behaviors. Those with this weakness, some believe, have an "addictive personality." It is their nature to be addicted easily and permanently.

Scientifically, however, this notion is unsupportable. At its core is a misapplication of several established, replicated research findings, including the following:

- Some people apparently are predisposed to developing addictive disorders such as alcoholism, amphetamine addiction, and dependence on the nicotine in tobacco. Even so, the predisposing factors differ between substances and life conditions, such that those predisposed toward alcoholism might not be predisposed to abuse cocaine or any other substance. This predisposition is not entirely ruled by genes, either. For instance, someone with a predisposition to alcoholism who has limited contact with alcohol might never develop the disorder.

- Substance-use disorders are more common in people with certain personality styles and personality disorders. However, no constellation of traits that could be identified as an "addictive personality" has ever been identified, despite major research efforts toward that end.

- Many behaviors can become compulsive, including such diverse actions as hand washing, counting holes in the tiles of ceilings, touching a certain object, gambling, performing mealtime rituals, cleaning doorknobs, and engaging in specific sexual practices. In short, most human behaviors can become compulsively driven, although they are not addictive. This is because, in some way or another, the behaviors are reinforced or are reinforcing. Some people may also eat compulsively, and some may engage in substance-*related* behaviors compulsively. This does not imply addiction, but rather may indicate some variant of a fixation or a compulsion, which are separate demons.

- Some people apparently are predisposed to developing compulsive behaviors. These are not necessarily the same people as those predisposed to developing alcoholism or drug abuse.

Despite heroic efforts and the expenditure of many research dollars, science has never been able to establish the existence of an addictive personality style that predisposes people to a variety of addictions and compulsions. Convincing the general public that this phenomenon does not exist may be difficult, since we have become accustomed to hearing about "chocoholics" and "foodaholics," and since the concept of an addictive personality offers a simplistic, intuitive explanation. In reality, the notion of the addictive personality remains unproven, undemonstrated, and most likely incorrect.

Has Science Ever Defined Addiction?

Well, yes and no.

Facts exist independent of who is citing them or whether the person is using them correctly; it is the way they are cited that frames their interpretation and results in misapplied meanings. Few areas of substance abuse debate are as riddled with ambiguous meanings and syntactical nuances as is the word *addiction*.

Some tobacco-financed researchers and industry supporters have asserted that one reason for not believing that nicotine is addictive is their claim that science has no adequate, consistent definition of the term *addiction*. This claim surfaces in such diverse places as peer-reviewed scientific journal articles, newspaper columns, litigation, and testimony before Congress. To the contrary, scientists with no financial ties to the tobacco industry have argued that the question is not one of defining addiction, but rather one of defining the properties of nicotine itself. Whether or not we use the word *addiction*, does nicotine have traits in common with substances that we commonly identify as addictive?

Some scientists defined *addiction* to their satisfaction several decades ago, but the word was appropriated into other contexts and stripped of its precise meaning. The term was first applied to chronic use of opiates, including morphine and heroin. Originally, addiction was identified primarily by compulsive use, physical dependence, tolerance, and damage to the user and society. Readministration of the substance was found to relieve abstinence-related withdrawal symptoms, leading to the concept of physical dependence as a "central defining characteristic of addiction," as explained by pharmacologist Caroline Cohen, writing with colleagues Wallace Pickworth and Jack Henningfield.

When the term *drug addiction* became associated with pejorative images, the World Health Organization recommended using the term *drug dependence*. As Cohen explained, drug dependence is a "psychic and sometimes also physical" state marked by a compulsion to use a given substance continuously or periodically to experience its effects and sometimes to avoid the discomfort brought on by its absence. This state may or may not include tolerance. People can develop both tolerance to drugs and physical dependence on drugs that are not abused. Similarly, patients in experimental settings do not seek out some drugs despite having developed tolerance and physical dependence.

Cohen clarified: "Addictive agents have one common attribute: the creation of a behavioral response repertoire often referred to as 'psychic dependence.'" While some would claim that "psychic" dependence differs from physical dependence, this view denies the fact that psychological dependence has a physiological basis.

When it became fashionable to label as addictive such diverse activities as compulsive gambling, overeating, and frequent sex, the word *addiction* lost its moorings. Nonetheless, the term still can be applied to any drug use that involves "drug-seeking behaviors," compulsive use of the substance, denial of the consequences of using the substance, and relapse after cessation.

The scientific and medical communities have moved away from the term *addiction* and shifted instead toward the term *dependence* as a way to describe serious substance-use disorders, including addiction to nicotine. *Dependence* is a precise term with specific published criteria and implications. To meet criteria for dependence on a substance, someone must have experienced at least three of these symptoms:

- Tolerance.
- Withdrawal when the substance isn't used.
- Using more of the substance than was intended, or using it over a longer period of time.
- Persistent desire or unsuccessful efforts to cut down or to control the use.
- Spending considerable time obtaining the substance or recovering from its effects. In the case of smoking, this could refer to spending a lot of time smoking, rather than engaging in tobacco-seeking activities. It is also possible that this criterion is not applicable to most use of nicotine.

These criteria, published in the fourth edition of the *Diagnostic and Statistical Manual* of the American Psychiatric Association and thus used for formal diagnosis, have become the basis for identifying and researching numerous substances of abuse, including hallucinogens, amphetamines, opiates, alcohol, and tobacco. Although they are codified, they are by no means universally accepted by the scientific community or applied in research paradigms. Some researchers choose to define addiction and dependence much more simply: the inability to stop a drug-reinforced behavior when one wants to stop.

The concepts of *dependence* and its less severe cousin *abuse* provide a useful heuristic for understanding what constitutes nicotine addiction. Although they have become a gold standard of sorts, they are not the final word. For example, The National Household Survey on Drug Abuse included four key questions to determine nicotine dependence, as part of a much larger survey about overall substance use:

- Current smokers were asked whether they had felt that they needed cigarettes or were dependent on cigarettes during the previous year.
- Smokers were asked whether they had needed more cigarettes to get the same effect.
- Smokers who had tried to cut back were asked whether they felt unable to do so.
- These same smokers also were asked whether they had experienced withdrawal symptoms or felt sick when they stopped smoking or cut down on cigarettes.

These questions were not a comprehensive measure of nicotine addiction and did not measure all symptoms of nicotine withdrawal. Thus they probably underestimated the proportion of smokers who would qualify as dependent. A 1989 committee reporting to the Royal Society and to Health and Welfare Canada defined drug addiction as this: "a strongly established pattern of behaviour characterized by (1) the repeated self-administration of a drug in amounts which reliably produce reinforcing psycho-active effects, and (2) great difficulty in achieving voluntary long-term cessation of such use, even when the user is strongly motivated to stop."

Cohen and her colleagues added to this definition of dependence their summary of compelling research evidence that smoking is "a highly controlled or compulsive behavior." The control in smoking comes from the precision with which smokers obtain nicotine to maintain consistent blood levels. They cited as proof the consistent patterns of cigarette smoking; the gradual increase of cigarette intake over time until a stable level is achieved; and the findings that more than three-fourths of current smokers say they would like to quit, and two-thirds have made at least one serious attempt.

Additional evidence of nicotine's addictive nature is that it is considered rewarding by smokers, even to the point of being what Cohen and her colleagues called a "potent euphoriant."

Addiction or Habit?

Calling smoking merely a "habit" troubles many scientists who study the effects of nicotine and tobacco. "One of my linguistic pet peeves is the use of the word 'habit' in reference to smoking and tobacco use," wrote Ronald Davis, former director of the U.S. Office on Smoking and Health, now editor of the journal *Tobacco Control*. He decried the repeated use by the tobacco industry of the 1964 Surgeon General's report to justify their claims that nicotine is merely a habit. The 1964 document concluded: "The tobacco habit should be characterized as an habituation rather than an addiction." However, Davis noted, "[T]he industry invariably skips over the preceding paragraph," in which the report stated that habitual use was "reinforced and perpetuated" by nicotine's action on the central nervous system.

The 1964 report did not label nicotine as addictive, referring to it instead as a substance used habitually, like coffee. Those insisting that tobacco isn't strictly addictive latched onto this distinction. A smoker's morning cigarette, they said, was no different than a morning cup of coffee. It must have been disconcerting for them to read reports in respected medical journals in the mid-1990s that caffeine can also be addictive. Quitting caffeine use when one routinely has even a small amount per day can result in abstinence symptoms that often include a nasty headache that can last for days. However, this physical dependence usually is not accompanied by compulsive drug-seeking behavior.

Unlike the 1964 report, the 1988 Surgeon General's report (subtitled *Nicotine Addiction*) was devoted almost entirely to nicotine's addictive qualities. With the benefit of hindsight, the latter report explained the nuances of terminology and classification that had resulted in the 1964 report, and stated that the terms *drug addiction* and *drug dependence* are equivalent.

The debate about whether or not nicotine is addictive has centered on several assertions, as reiterated by R. J. Reynolds researchers John Robinson and Walter Pritchard, who have participated in scientific forums. Robinson and Pritchard argued:

- The scientific community lacks a precise definition of addiction, "apparently using the word to indicate any behavior that people engage in and may find difficult to stop."
- Classifying a drug or behavior as addictive "because some people may find it difficult to stop" is not in the best interest of scientific inquiry.
- Nicotine use can be seen as a habit, or as habituating (not referring, incidentally, to the precise meaning of that term as applied in the behavioral sciences).
- Because the use profile of nicotine differs from that of known addictive substances such as heroin, it should not be classed with them.
- Because nicotine is not "intoxicating" and does not impair motor performance (i.e., tasks requiring motor skills) or perception, it does not fit scientific criteria for addiction as established in the 1960s through 1980s.
- Nicotine has no "strong" euphoriant effect such as that of cocaine, although its use is "pleasurable."

- Research with laboratory animals suggests that nicotine self-administration in nonhumans is neither readily established nor robust. (Such self-administration is a standard research technique for studying drugs of abuse such as cocaine or heroin.)
- Since the "non-pharmacological aspects of smoking," such as how the smoke tastes, appear to drive smoking behavior, smoking is "more accurately classified as habit than addiction." This is true, they claimed, because smoking is "a complex behavioral process involving both pharmacologic and non-pharmacologic factors."

Just how accurate are the tobacco industry's claims? Many scientists and policymakers are skeptical of arguments emanating from researchers on the payrolls of the tobacco industry, even if the arguments have validity and serve to nudge science toward a clearer definition of addiction. The 1988 Surgeon General's report compared nicotine addiction to that of the "hard" drugs, including the opiate heroin. It concluded that tobacco does have many addictive properties in common with other drugs of abuse. As a toxic substance, it often makes first-time users queasy, if not actually nauseated. With continued exposure, users develop a tolerance and can use increasingly larger quantities. When smokers are deprived of nicotine, they may experience measurable, unpleasant withdrawal symptoms. These characteristics, true of many hard drugs, are also well documented for nicotine use.

A small percentage of tobacco users continue to use nicotine at a low level for many years without developing tolerance or dependence. Their personalities, lifestyles, and metabolism of nicotine resemble those of smokers who are nicotine dependent. These low users are not naïve smokers, since they consume many thousands of cigarettes over the years that they smoke. However, they are anomalous smokers. They follow a daily use pattern that remains minimal, and they experience virtually no abstinence effects when they quit using tobacco.

Is the existence of such smokers, as described by Saul Shiffman in 1989, a valid argument against characterizing nicotine as an addictive substance? If nicotine were addictive, according to those arguing against that idea, shouldn't it be addictive for all users? That argument actually works against its proponents; the term *chippers,* which is often used to describe these low-use smokers, originated with the scientific and medical research on opiate use. The existence of chippers among opiate users does not call the addictiveness of opiates into question.

But staking the determination of nicotine's addictiveness on the slippery definition of *addiction* seems to beg the question. The imprecision of one term does not change the nature of the substance. No one has claimed that tobacco produces the same "high" as cocaine or the same intoxication as marijuana, at least in the doses commonly used by cigarette smokers. A more fundamental question is whether nicotine fits the profile of drugs known to be addictive, or carries a high risk for leading to chronic use. Thus, the questions addressed by the medical and mental health community are these: Do people become tolerant to higher and higher amounts of nicotine delivered over the course of the day? Do regular users of nicotine experience abstinence effects when they go without nicotine? If the answer is an overwhelming yes, it may indicate a physical dependence. Perhaps it matters little that common usage refers to people being "addicted" to Twinkies, or even that many tobacco-related behaviors can be labeled as habitual.

The nature of addiction is that each substance with a potential for abuse, including nicotine, has a unique profile of use. Each substance requires separate, specific examination. Why should the effects of inhaling nicotine have to be virtually identical to the effects of injecting heroin for the similarities to be worth noting? The commonalities provide evidence that both substances can be used deleteriously; the differences merely illuminate the complex nature of human addiction.

Nicotine in the Brain

From the first cigarette, smoking changes the brain. The brain's billions of nerve cells communicate through the electrical and chemical activity of substances called *neurotransmitters.* Nerve impulses travel as electrical signals and are transformed into chemical signals. Neurologist Richard Restak described neurotransmitters as being like ferry boats "steaming across a channel toward the 'loading dock,' the receptor on the membrane of the receiver cell."

The receptor cells, or Restak's "loading dock" cells, can increase in number and in activity. When a person uses a large amount of a certain substance (such as nicotine), the number of receptors for this substance increases. Restak described it metaphorically as "the basis for the withdrawal response in addiction when the receptors, deprived of their usual supply of addicting substance, 'cry out' like deserted lovers for the missing chemical." Without the chemical for a long enough period of time, the receptors are "down regulated." The circuits of the brain adapt to the presence of nicotine, whose structure mimics the structure of a naturally occurring brain chemical called *acetylcholine*, a substance that releases another substance called *dopamine*.

Molecules of dopamine are released by the brain when pleasurable events occur, such as the everyday enjoyment of petting a cat or eating pizza. Some drugs that alter mood, including nicotine, trigger dopamine in the same way that life's big and little pleasures do. Brain cells adapt to the unnatural presence of these substances by changing both the sensitivity of receptors and the number of receptors.

Quite a different series of events occurs when smokers attempt to quit. Without the customary dosage of nicotine, the brain triggers several reactions that are normally associated with negative experiences such as punishment. As the lower blood levels of nicotine result in reduced dopamine to the brain, smokers experience withdrawal (or abstinence) symptoms, including irritability, anxiety, frustration, and depression. All that is necessary for the negative feelings to flee is for the smoker to replenish the body's supply of nicotine.

Pharmacologists explain the addictive nature of drugs as a reflection of the ability of the substance to enhance the transmission of dopamine at specific sites in the brain. The neurotransmission-enhancing effects of some nonnicotine substances are so strong, for example, that they alone can explain the addictive properties of the substances.

Nicotine is known to evoke an increase in "dopaminergic overflow" in a portion of the brain where neurotransmission-enhancing effects occur. However, not all smokers smoke alike. British researcher Michael Russell, in a 1990 report, identified at least two distinct types of smokers, "peak seekers" and "trough maintainers." The peak seekers are those who smoke to achieve a substantial peak nicotine level after each cigarette. This most likely stimulates the nervous system's central *nicotinic receptors*. Trough maintainers, on the other hand, smoke more frequently to maintain a relatively constant nicotine level, which results in the same receptors being blockaded so that the "loading dock" cells are closed to incoming transmissions. Peak seekers' blood levels of nicotine dip and rise dramatically over the course of a day, perhaps resulting in repeated stimulation of a dopamine system. Trough maintainers' blood nicotine levels are fairly constant throughout the day. As Scottish scientist David J. K. Balfour explained: "It is possible . . . that [smokers] adjust the way in which they smoke so that [they achieve] the appropriate combination of nicotinic receptor stimulation and desensitization which they find most rewarding." Desensitization refers to a process of diminished responding to a repeated stimulus, akin to the common situation of growing used to something that initially triggered a response.

The stimulation and desensitization processes Balfour described seem to occur commonly in smokers and may be part of why nicotine is addictive. Some scientists speculate that the desensitization effect may be a component of the anxiety-relieving, or *anxiolytic*, effects documented in tobacco research and evident in so many smokers who turn to nicotine to relieve stress. If this is so, desensitization may involve a different neural mechanism than the one activated by prescription drugs such as Valium (diazepam).

Is Addiction a Disease?

It was not long ago that alcoholism was blamed on lack of character. And it was not long ago that smokers believed they could quit if only they had enough will power. Certainly, strength of character and

Chipping Away

fewer than 6	number of cigarettes smoked per day by the typical tobacco chipper
none	withdrawal symptoms experienced by chippers when they quit smoking
8.2%	Australian smokers who are chippers
15%	California smokers who are chippers
decades	how long the typical tobacco chipper smokes
tens of thousands	total cigarettes consumed over a chipper's lifetime of smoking
none	cigarettes per day considered safe to smoke

strong will can boost attempts to overcome an addiction—or a habit, for that matter. Nevertheless, our present understanding of the biology of nicotine addiction offers both clarity and charity. In the last several decades, science has brought some humanity and compassion to our understanding of the conundrum of addiction by showing that human vulnerability to numerous drugs of abuse is widespread. Defining a substance as addictive and defining its overuse as a disorder or a disease have helped free users from the trap of labels. Some people argue that in using a disease model for understanding addictions, we have given substance users a convenient excuse for not changing behavior by hiding behind arguments that their problem is "genetic" or is out of their control. While this claim may be accurate in part, it does not take into account the benefits that may come from biologically based explanations of addiction.

The first addiction to be recognized as a disease was alcoholism. In view of that, some have asked whether nicotine addiction is also a disease. Addressing this question requires that we consider the definition of *disease*. Whether or not nicotine addiction can be termed a disease depends on the answers to several questions. Is nicotine use under the control of the tobacco user? If nicotine use is addictive, then the onus is outside the individual smoker, even though the initial trials with cigarettes were the smoker's choice. As clinical researcher Norman Miller explained: "The primary foundation for considering nicotine addiction to be a disease rests on the acceptance of the loss of control by the nicotine addict."

Attributing the undesirable consequences of the addiction to "a disease concept" removes the weight of viewing smoking as a "moral dilemma" and thus facilitates the earliest steps toward cessation. Miller added: "[I]nsistence on correcting a weak character or treating an underlying psychiatric or emotional disorder will not initiate abstinence or prevent relapse to nicotine. Nicotine addicts are already filled with self-condemnation, and a further exaggeration of the guilt by making the addicts at fault for their smoking will further impede the addicts' accepting responsibility for treatment of the nicotine addiction and its consequences."

When Russell asked heroin users to rate drugs in terms of being "needed," they put cigarettes at the top of their list. They said that they perceived coping without cigarettes to be more difficult than coping without heroin. Russell concluded in a 1990 report that "cigarette withdrawal is no less difficult to achieve and sustain than is abstinence from heroin or alcohol." Additional evidence of nicotine's powerful pull is in the statistic that between 45 and 70 percent of smokers who survive a heart attack resume smoking again within a year. About half of all smokers who undergo lung cancer surgery take up smoking again.

Where the Argument Leads

Few scientists, and only a small percentage of smokers, doubt that nicotine is addictive, by some definition. Much is at stake in the addiction debate. The risk of the rhetoric is that the realities of tobacco use and the smoking experience will be lost in a terminology war. Discussions over the definition of addiction can become a smoke screen obscuring the more fundamental and burning question: What makes people smoke?

The question of addiction is at the heart of initiatives by the U.S. Food and Drug Administration (FDA) to exert more government controls over the manufacture and sale of cigarettes and other tobacco products. Former FDA commissioner David Kessler testified in unequivocal language that between 74 percent and 90 percent of all smokers were addicted. He summarized in 1994: "Accumulating evidence suggests that cigarette manufacturers may intend this result—that they may be controlling smokers' choice by controlling the levels of nicotine in their products in a manner that creates and sustains an addiction in the vast majority of smokers. . . . Whether it is a choice by cigarette companies to maintain addictive levels of nicotine in their cigarettes, rather than a choice by consumers to continue smoking, that in the end is driving the demand for cigarettes in this country."

Kessler recounted how the one-time "simple agricultural commodity" of tobacco eventually evolved into the production and marketing industry for a "nicotine delivery system." To the surprise of many who heard his testimony, he explained that by reconstituting tobacco stems, scraps, and dust, cigarette makers began controlling and manipulating nicotine levels to achieve maximum addictive potential. He showed charts of patents in which tobacco companies added nicotine to tobacco rods, filters, and wrappers. Other patents indicated control of nicotine levels by extraction and utilization of new chemicals. Kessler stated: "Patents not only describe a specific invention. They also speak to the industry's capabilities, to its research, and provide insight into what it may be attempting to achieve with its products."

He cited the industry's achievements:

- Controlling the amount of nicotine to provide a desired psychological effect
- Increasing the amount of nicotine by manipulating nicotine levels
- Controlling the rate at which nicotine is delivered
- Transferring nicotine from one material to another
- Adding nicotine to other parts of the cigarette.

"Since the technology apparently exists to reduce nicotine in cigarettes to insignificant levels," Kessler asked, "why . . . does the industry keep nicotine in cigarettes at all?" And, similarly, "With all the apparent advances in technology, why do the nicotine levels found in the vast majority of cigarettes remain at addictive levels?" If nicotine is merely a flavorant, producing a burning in the throat to which smokers become accustomed, "why not use a substitute ingredient with comparable flavor, but without the addictive potential?"

Why not, indeed.

So, Is Nicotine Addictive?

On this question, some argue that the jury is still out. Others argue that the jury has never been presented with an adequate case. Others believe that the verdict was sealed more than a decade ago.

Here is a suggestion for those who aren't yet convinced. This winter, some day when it's 20 degrees Fahrenheit outside and snowing, look outside an office building in which smoking is prohibited. Note the smokers huddled outside in the doorways, braced against the weather. Ask yourself: Are they out there just for the pleasurable sensation of smoking?

CHAPTER 5

I sip on the slushy ice in the bottom of my soft-drink cup. She says
she doesn't need anything to drink right now.

"I don't drink coffee," she shakes her head. "That's another reason
why it's very hard for me to quit smoking, because the nicotine is a
stimulant. I don't drink coffee because I *can't* drink coffee, and I've
never trained my body to be addicted to caffeine. A very little amount
of caffeine can send me into hyperspace. It affects me very strongly. I
can do small amounts of tea."

I ask how this affects her ability to stay up late as a theater stage
manager, which she does part-time.

"If I'm doing a show, and I have to be up until two or three in the
morning, the nicotine is what keeps me awake," she explains. "The
nicotine—that's my coffee. I don't ever use it to help me go to sleep. I
used to, I tried it a few times, before I realized that I would just end up
more awake afterwards.

"Some people say they use a cigarette to help them relax. Well, it doesn't help me relax. It helps me to focus my thoughts. I do notice that sometimes, not most of the time, but in very rare situations, when I'm stressed, then I will smoke faster. When I'm out drinking, when I'm in a bar with friends, then I can smoke more slowly."

Steadying the Ark

We walk the streets and share the stairwells with them. In the winters, we catch their colds, or perhaps give them ours. From the other side of the elevator, we hear the music that leaks from the edges of their headphones. We may have been where they are, and returned. Or perhaps we may yet be among them. Many of us will yet learn to carry the burden of a psychiatric disorder.

Evidence is mounting that people with a variety of psychiatric problems such as depression, hyperactivity, anxiety, and eating disorders are at increased risk for smoking. Additionally, the high rate of smoking among patients with schizophrenia is well established. Persons with these disorders may need to get professional treatment for the disorder before they can successfully quit smoking. This also may mean that researchers will need to develop new smoking treatments tailored for persons with specific psychiatric problems.

The rest of us who smoke may have little notion of the extra problems faced by smokers whose smoking is tied to a psychiatric problem. For them, quitting tobacco may involve nothing more than enduring withdrawal symptoms and adjusting behaviors; it may mean the exacerbation of their psychiatric symptoms, such that what is now a manageable condition can worsen and become unpredictable. The nicotine in the bodily system of a smoker affects how some psychiatric medications work. When the smoker stops using nicotine, the medications work less effectively or predictably. Whatever symptoms nicotine has affected, either directly or indirectly, can change.

Also, smokers whose symptoms have been masked or moderated by nicotine may find that quitting smoking causes symptoms to appear. A smoker prone to depression may find that quitting smoking results in the recurrence of depressive symptoms. A person with an anxiety disorder may find that no longer using nicotine results in a groundswell of anxiety symptoms. A smoker with schizophrenia may find that her antipsychotic medication becomes more potent when she quits smoking. Some schizophrenia symptoms may suddenly worsen when a smoker with schizophrenia quits. Many smokers who try to quit find the experience noxious, uncontrollable, and unpredictable. They are not sure what is happening inside them, or why. As a result, they continue to smoke.

Smoking is commonplace among people with several psychiatric conditions, including depression, schizophrenia, and substance abuse. Patients in treatment for these problems can quit smoking without serious impact on their ongoing treatment if their medications are managed within the framework of smoking cessation. The most disturbing symptoms of their psychosis or their depression could remain controlled. They do not need tobacco to stay rational or to feel healthy.

Nicotine produces a range of effects on human emotions. Scientists have proposed various neural and physiological mechanisms for nicotine's effects, which vary according to the place in the brain where nicotine is absorbed, the amount that is absorbed, and the rate of absorption. In addition, a smoker's state, such as being agitated, rested, active, or quiescent, determines the effect of nicotine. Depending on how it is taken into the body, nicotine can be either sedating or stimulating. Used in subtly different ways, it can have dramatically different effects; it can help an anxious person relax and can provide an emotional lift for a person who is depressed.

Mood and Nicotine

Evidence abounds that tobacco use affects mood. Tobacco companies themselves tout the "pleasure" of smoking, not just in sensation, but in emotional effect. The reality is that even if smokers don't necessarily like the *fact* that they smoke, they seem to like to smoke. Nicotine may not be an intoxicant, but it does appear to be a *euphoriant*, or a substance providing feelings of euphoria. The research team of Cynthia and Ovide Pomerleau of the University of Michigan demonstrated this in a 1992 study showing that different amounts of nicotine produced varying euphoriant effects. Smokers reported fewer euphoriant sensations when they used an ultra-low-nicotine cigarette rather than a medium-nicotine cigarette or their usual brand. When subjects smoked a high-nicotine cigarette, they reported at least one euphoria sensation that started about 2.5 minutes after the cigarette was lit. The Pomerleaus suggested that the effects, documented in relation to the subjects' blood plasma levels of nicotine, reflected what is called a *dose-response* relationship. In other words, the higher the level of nicotine, the more pronounced the euphoria, at least to a point. The phenomenon was particularly observable when subjects had been abstinent from tobacco overnight. It was also more pronounced in the smokers reporting a greater dependence on nicotine.

Scientists have been aware for decades of what is called nicotine's *paradoxical effects* that vary according to dose and means of delivery. At the low doses of cigarette smoking, nicotine appears to be a likable substance in those who have used it long enough not to be affected by its toxic effects. Although cigarettes are not rated as pleasurable as alcohol or other drugs, nicotine rates well on "drug-liking" and euphoria rating scales. In these studies, a subject is given nicotine without knowing which of several substances it might be, or when it will be administered, and then is asked to report how much he or she likes it, or to note when he or she feels euphoric. Other research has shown that intravenous infusion of nicotine produces a "rush" somewhat similar to that of cocaine or morphine, although shorter in duration. Simi-

Nothing to Smile About

2.9%   nonsmokers who have experienced serious depression

6.6%   smokers who have experienced serious depression

25% – 50%   patients seeking help quitting smoking who have a history of serious depression

28%   smokers without history of depression who are likely to quit smoking

14%   smokers *with* history of depression who are likely to quit smoking

52%   people without history of depression who have ever smoked

76%   people *with* history of depression who have ever smoked

30%   smokers without history of depression who report feeling depressed in the first week after quitting smoking

75%   smokers *with* history of depression who report feeling depressed during the first week after quitting smoking

larly, the administration of a drug that blocks nicotine's effects, the drug antagonist mecamylamine, also attenuates subjects' nicotine-liking scores.

Fundamental to understanding these effects is a comprehension of the overall anatomy of the brain, which is shaped somewhat like the meat of a walnut. Its two halves are connected by several bundles of fibers that allow the two sides, with their respective specialized functions, to communicate. The brain functions, however, as a unitary whole, not as two competing sides (left-brain/right-brain theories and arguments with oneself notwithstanding). Although certain tasks, such as comprehending written language, are relegated to specific parts of one side of the brain, the information is rapidly transmitted to other areas of the brain as well. Streams of information continually travel throughout the brain, needing neither a passport nor luggage to go from one side to the other. The intact brain is, indeed, one organ.

Nonetheless, parts of the brain are specialized. As the brain develops before birth, through early childhood, and into adolescence, cells migrate to predetermined locations throughout the brain, developing into intricate structures with distinct functions. Not only are functions focused to allow primary processing in specific areas of the brain, but the two hemispheres of the brain approach information processing differently. The right hemisphere, as a whole, tends to process information more holistically and conceptually. The left hemisphere tends toward a more step-by-step, analytic interpretation. If our brain is intact, we rely on neither one side nor the other, but rather on the interaction and synthesis of both. Having two sides to our brains not only affords us a developmental spare in case one side is damaged, but also allows us a richness of interpretation, as information is analyzed in different but complementary ways.

Emotions are believed by many scientists to be based in what is called a *lateralized* function, a function that is mediated primarily on one side of the brain. While emotion does not originate entirely in either hemisphere of the brain, studies of localization of emotional processing in the brain have shown that the hemispheres play different roles in our emotional comprehension and expression. We read others' emotions with parts of our brain that, for most people, are located in the right hemisphere. We examine what we have read with input from the left hemisphere. We apparently use right-hemisphere capacities to generate emotionally expressive output, although without the left hemisphere, that output might be limited to mere sound and fury.

One relatively noninvasive, inexpensive way of watching the brain work is with the technology of the electroencephalograph, or EEG. This procedure monitors and allows measurement of the brain's electrical activity. A dense array of EEG electrodes located all over the scalp can be mathematically and visually transformed to provide a picture of the brain's functioning. EEG and related research and diagnostic techniques provide a way of observing the brain's electrochemical activity in process.

EEG studies of the brain's response to the administration of nicotine show a fairly consistent picture. Nicotine use is associated with relatively greater right-hemisphere EEG changes than left-hemisphere changes. Since the brain's neurotransmitter systems are somewhat different in the two hemispheres, and since a given drug binds to receptor sites according to patterns specific to one side of the brain or the other, the finding that nicotine does not affect both sides of the brain in the same way is no surprise. As for what this implies, that issue may be best addressed by looking at nicotine in other contexts.

Nicotine's effects may be linked to a person's state when using the substance. For example, when research subjects are relatively unaroused (referring here to alertness and vigilance, not to matters sexual), a modest dose of nicotine makes them more alert and increases the activation of their brain's cortex, or outer layer of gray matter.

However, nonhuman animals given nicotine when they are aroused or stressed experience a decrease in certain EEG measures and in behavioral signs of arousal. This raises the question, does nicotine act as a modulator of emotions? The answer is less clear when the question is considered in humans. Apparently, the effects of nicotine vary not only according to one's state of arousal, but also according to the complexity of one's emotional and arousal state.

Oddities and Entities

"We must not overlook the fact that many of the young gentlemen who smoke in America have other unhealthy habits, and their mode of living is often demoralizing."

British Medical Journal, 1890

Bears for Butts

California campaign countering the assertion by a tobacco executive that cigarettes are no more addictive than Gummy Bears; anyone turning in cigarettes got a package of Gummy Bears

"I love cigars . . . .
I love them for the same reason I love being an Episcopalian."

India Allen, 1988 Playboy Playmate of the Year, quoted in *Cigar Aficionado*

"Smoking is one of the leading causes of statistics."

Attributed to writer Fletcher Knebel

"I assume you light the color-coded end, right?"

Dilbert's co-worker Wally, upon taking up smoking so that he could take more work breaks

---

Some studies do show that nicotine has a dampening effect when used during a state of heightened arousal. Other studies show that nicotine increases arousal even when the user is already highly aroused. Yet other studies show that nicotine's effects depend on the baseline rate of behavior. These varied findings may reflect differences in research paradigms, and the difficulty of studying such subtle and transient things as emotions.

In contrast to all those models, nicotine researcher David Gilbert proposed that smoking can lead to "simultaneous increases in subjective alertness" as well as to increased EEG arousal, *and* can still decrease stress. He and others have concluded that stress and arousal may be "orthogonal dimensions," or separate entities.

Gilbert reported in a 1995 article that extroverted smokers responded to smoking with diminished drowsiness and decreased slower-wave EEG activity. (The slower our EEG waves, the drowsier and less alert we feel.) Smokers testing as more neurotic and depressed responded to nicotine with increased drowsiness and slower-wave EEG activity. Higher rates of baseline depression symptoms were associated with right-hemisphere EEG activity in the frontal lobes of the brain, a pattern characteristic of depressed persons. Smoking normalized this left-right asymmetry between the hemispheres of the brain.

The complexity of nicotine use is obvious in these varying ways of examining the relationship between mood and tobacco use. To some, it may seem that these multiple approaches are analogous to the tale of the blind men describing an elephant. The important fact is not that the descriptions illustrate different aspects of the elephant, but that the elephant is indeed there. However it is described, the connection between nicotine use and emotional functioning does appear to be as real as an elephant, and for many smokers it looms as large.

Depression and Nicotine

Depression is everywhere throughout the world. About 3 percent of those in the United States will, sometime in their life, experience a depression severe enough to be classed as a Major Depressive Disorder, using the nomenclature of mental health professionals. To meet criteria for such an episode, a depression must involve more than merely feeling down or blue. Someone whose sleep patterns are disturbed, who has lost interest in activities that used to be pleasurable, or who has gained or lost weight without consciously trying may be depressed. Someone who feels hopeless, who wishes he or she were dead, or has planned suicide may be severely depressed, and most likely could benefit from professional help.

As distressing and serious as these symptoms sound, they are well documented and remarkably similar in cultures all over the world. Women tend to experience them with greater prevalence than men. Children and adolescents experience them. Good people and bad people experience them. Research suggests that some people are genetically predisposed to developing depression, although many internal and external factors apparently can trigger depression. For example, some women experience depression while pregnant, others experience it after having a baby, and yet others are depressed at not being able to have a baby. Sometimes depression seems to focus on an issue, but just as often it does not. Depression is treatable by medication, sometimes by psychotherapy, sometimes both; and sometimes it stubbornly persists until it spontaneously wears itself out. For some people, it never abates.

Nor is it always evident at diagnosable, clinically relevant levels. Many people live out their lives with a subclinical level of depression as nagging and wearing as a persistent infection or a low fever. This dysphoria (also termed *dysthymia*) casts a subtle gloom over their lives. For those who are chronically depressed, battling varying degrees of dysphoria becomes the struggle of a lifetime. The current widespread availability of antidepressants is a godsend for many depressed persons, although others are neither helped nor satisfied by medications.

Some people find that potentially dangerous and addictive substances offer relief from the flatness and bleakness of depression. Often, that relief becomes a trap.

Enter tobacco.

Psychiatry research articles published several years ago began to raise the possibility that depression might be a cause of smoking, although the research did not contradict the notion that depression might have a role in maintaining smoking. Naomi Breslau and colleagues Marlyne Kilbey and Patricia Andreski examined whether smokers with a history of depression were at increased risk for developing severe nicotine dependence. Conversely, they asked in a 1993 paper, were those with a history of nicotine dependence at greater risk for depression? The answers, based on interviews with young adults, were *yes* and *yes*. This could imply causality, they said, but more likely it pointed to a mutual cause of both smoking and depression.

Breslau also noted in a 1995 report that dependent smokers had increased odds for alcohol and illegal drug use disorders, for major depression, and for anxiety disorders. The presence of major depression and an anxiety disorder were both associated with nicotine dependence. Persons with a history of conduct problems in youth had increased odds for developing nicotine dependence. She explained that smoking and depression are "bi-directional," in that both conditions increase the odds for the onset of the other. This suggested that both were caused by some as-yet-unidentified common predisposition, such as neuroticism, but that neither smoking nor depression causes the other.

A 1998 article by Breslau and her colleagues concluded that depression contributes to daily smoking across the life span, and that the influence of depression on smoking begins in adolescence. The investigators followed more than 1,000 young adult smokers for five years to collect the data reported in *Archives of General Psychiatry*. Their data did not support the notion that depression led to the initiation of smoking by their subjects. They did suggest, however, that attempts to "self-medicate depressed mood" could explain the subjects' movement toward daily smoking.

The researchers also indicated that smokers were at greater risk for developing major depression, but they did not attribute this to smoking. Rather, they looked toward what they termed "shared etiologies," or causes that both depression and smoking have in common, such as social environment, personality, and coping styles. They explained that early conduct problems appeared to be related to both smoking and depression, and that these associations accounted in part for the connection between depression and smoking.

Their study contradicted earlier cross-sectional reports that depression made smoking cessation more difficult. Instead, Breslau and her co-authors found no indication that depression influenced cessation.

Kenneth Kendler and his colleagues worked with a group of twins to evaluate whether smoking and depression are causally related. After controlling for factors such as personal and family smoking and depression history, they concluded that the relationship between smoking and depression occurred "solely" from familial factors that predisposed people to both conditions. In women, they explained, the relationship between smoking and depression was not causal but came primarily from factors that could be genetic, related to family environment, or both. Their study was the first published report to document a strong association between smoking and future depressive episodes.

The risk for suicide also has been associated with smoking. A physician at the University of California at Davis, Bruce Leistikow, combined numerous studies into one overall analysis to identify a link between suicide and smoking. Previous research had shown that smokers had elevated rates of suicidal thoughts, attempts, and completions. As with the findings linking depression and smoking, Leistikow's meta-analysis did not indicate that smoking causes suicide, but did demonstrate a significant connection. In published studies, higher rates of smoking have been associated with higher rates of suicide. Former smokers had lower suicide rates than current smokers.

Such links between emotions and nicotine use have other manifestations as well. Smokers who report high levels of negative emotions such as depression are less likely to quit using tobacco. High levels of such feelings also can statistically predict smoking, as well as the use of other substances of possible abuse. Smokers accurately expect that using tobacco will curb their negative emotional feelings. Not only that, but negative emotions can trigger a relapse to smoking among ex-smokers; the more negative the emotion, the greater the likelihood of actually relapsing, rather than just being tempted.

Even if depression doesn't necessarily lead to smoking, or vice versa, evidence is plentiful that the two do co-occur with some frequency. Researcher Sharon Hall and colleagues, examining this intertwining, explained in 1993 that smoking is perceived as reducing negative emotional state and stress, and enhancing mood. Nicotine also is a means of focusing attention away from disturbing stimuli or interference, and providing stimulation during boredom or inactivity. Similarly, nicotine withdrawal can result in negative, unpleasant feelings that peak one to two weeks after cessation and return to a baseline state about one month after the smoker stops. The uncomfortable sensations of depression, anxiety, restlessness, nervousness, irritability, impatience, anger, and aggression experienced by some quitting smokers may be more than just withdrawal symptoms: They may be "the emergence of preexisting psychopathology." In other words, smoking may help curb those symptoms; but when the smoking stops, the symptoms return.

Depressed people tend to engage in fewer pleasurable activities than do nondepressed people. They also experience more relationship problems, are less inclined to anticipate positive events, are more inclined to anticipate negative events, and may experience distorted thinking when they feel at their worst. As a result, they also may have less contact with the people, circumstances, and beliefs that might help them lift the depression. In this condition, the quick relief obtained from nicotine's mild euphoriant and anxiety-reducing effects is highly reinforcing. As Hall and colleagues pointed out, "Pharmacologically, it stimulates the organism."

No wonder so many of us organisms keep smoking.

Schizophrenia and Nicotine

Persons with schizophrenia are more likely than not to be smokers. Between 50 and 90 percent (the rate varies, depending on diagnostic criteria) of those with this debilitating psychiatric condition smoke. Unlike the common misperception that schizophrenia refers to having a dual personality, this disorder actually relates to psychotic or delusional thoughts, social withdrawal, and inability to function adequately in daily life. Untreated, those with schizophrenia might exhibit bizarre behavior and might be unable to experience pleasure. Two questions relate to the high prevalence of smoking among those with schizophrenia: Why do the majority of them smoke, and what happens when they stop?

As it turns out, many persons with schizophrenia not only smoke, they sometimes smoke to excess. Some patients, according to a 1986 report by Darrell G. Kirch and colleagues, used tobacco to the point of nausea, and developed water intoxication from reduced secretion of an antidiuretic hormone as a result of the excessive nicotine use. One reason for this overuse may be that nicotine appears to normalize a mental "gating" mechanism that typically malfunctions in those with schizophrenia. Lawrence E. Adler and his associates at the University of Colorado Health Sciences Center reported in a series of articles in the early 1990s that the administration of nicotine temporarily helped correct a sensory gating abnormality in which those with schizophrenia are hypervigilant in responding to sensory input that a person with normal functioning would ignore, such as background traffic noise. The administration of nicotine normalized a brainwave indicator of this gating defect in both smokers and nonsmokers. Nicotine temporarily and briefly overrode the brain defect that is characteristic of schizophrenia, thus giving patients a brief period of calm.

The inability of many persons with schizophrenia to filter out irrelevant stimuli apparently is an inherited trait that predisposes them to developing schizophrenia. The brain defect is not, by itself, sufficient to cause schizophrenia. Healthy siblings of some patients with schizophrenia also have the inherited trait, but they do not have schizophrenia. The gene that appears responsible for this condition is linked to a brain receptor that filters incoming information. Nicotine stimulates that brain receptor. Therefore, schizophrenic persons who use tobacco are briefly turning on this receptor and thus getting temporary respite from the information overload common to their condition.

This landmark genetic work is described in reports by Sherry Leonard and Robert Freedman at the University of Colorado School of Medicine and their colleagues. Scientists have not yet found the gene mutation that causes this condition; nonetheless, the gene discovery itself opens the possibility of new treatments that target this receptor.

Another clue as to the effects of nicotine in patients with schizophrenia comes from a finding that smokers using antipsychotic medication experienced less of a disorder called *akathesia*, a restlessness and inability to sit down that is a side effect of the medication, than did nonsmokers using the same medication. Schizophrenic patients who smoked experienced fewer involuntary movements and reported less jitteriness over the course of a four-week study in which they used fixed doses of antipsychotic medication. Using the medication resulted in increased involuntary movements in nonsmoking patients with schizophrenia.

The researcher conducting the study, Nicholas Caskey, hypothesized that people with schizophrenia may smoke partially to reduce the unpleasant side effects of antipsychotic medication. Nicotine may have some beneficial effect in lowering some of the side effects, particularly akathesia and abnormal movements. Even if this turns out to be the case, patients with schizophrenia would be better off using a less dangerous source of nicotine than cigarettes (for example, nicotine gum or patch).

Smokers with schizophrenia do find that they can quit using tobacco without an increase in their psychiatric symptoms. However, quitting tobacco might worsen a movement disorder called *tardive dyskinesia*, which is also a side effect of some antipsychotic medications.

Caffeine, Alcohol, and Nicotine

Caffeine, alcohol, and tobacco are the drugs most commonly used throughout the world. They are also commonly used together.

Most smokers are used to having coffee and tobacco together. In fact, smokers drink more coffee than nonsmokers. Exactly why this is the case isn't yet clear, although scientists have found some curious relationships between caffeine and nicotine. A smoker who also uses caffeine (which is to say, the majority of smokers) might find it easier to strategize quitting smoking if he or she understands that the two are connected.

The relationship between caffeine use and nicotine consumption is probably at least partially behavioral. Coffee consumption and ciga-

High Anxiety

77%	nonsmokers who drink coffee
86%	smokers who drink coffee
5 hours	average half-life of caffeine in the body of a nonsmoker
3.5 hours	average half-life of caffeine in the body of a smoker
50% – 60%	extent to which quitting smoking increases caffeine levels in the body by affecting how the body metabolizes caffeine
nervous	what heavy drinkers of caffeinated beverages feel when they quit smoking
anxious	ditto
jittery	ditto

rette smoking are so frequently done together that either can serve as a cue that triggers the other, but smoking most often follows coffee consumption, not vice versa.

A common criticism of research studies of common life events such as coffee drinking is that the results are not necessarily generalizable beyond the laboratory. Ken Perkins at the University of Pittsburgh has worked around this limitation by having smokers use both nicotine and caffeine under conditions comparable to those in which they would normally be used. Studying research participants in settings of rest and casual physical activity, he reported in 1994 that the cardiovascular and self-reported subjective effects of nicotine and caffeine may be additive.

Several factors could account for the frequent combined use of nicotine and caffeine. One could offset the other, since caffeine can increase anxiety and arousal and nicotine can decrease both. It is also possible that stress or the consumption of alcohol could contribute to the relationship. Whatever the cause of the connection, severing the tie could result in unpleasant effects. Caffeine is metabolized faster when it is consumed by someone who also uses nicotine. When a smoker stops using nicotine, caffeine is metabolized more slowly, and thus remains in the body longer. If caffeine consumption levels are unchanged, this results in a greater accumulation of caffeine in the body, with the pattern of elevation lasting for as long as six months.

Just as it is common for a smoker to sip a cup of coffee before using tobacco, it is also common for smoking to accompany the use of alcohol. Nicotine and alcohol have separate and distinct effects, both emotional and physiological. Perkins and his colleagues reported in 1996 how men and women respond to nicotine and alcohol separately and together. Nicotine, taken by itself, increases a feeling of "head rush," dizziness, jitteriness, tension, arousal, decreased fatigue, and relaxation. Alcohol also brings feelings of "head rush," jitteriness, and dizziness, but adds feelings of intoxication. It has no other stimulant effects. Taken together, nicotine attenuates some of the sedating and intoxicating effects of alcohol, a finding that had been reported in several previous studies utilizing performance and EEG measures. Perkins and his colleagues also found that men and women responded differently to the combination of nicotine and alcohol. Men using both substances together experienced a reduction in dizziness, relaxation, tension, vigor, and arousal. In contrast, women using both substances experienced enhanced effects of both substances.

Smoking cessation was accompanied by an increase in wine consumption in a large group of male World War II veteran twins studied by investigators Dorit Carmelli and associates at SRI International in California. Following alcohol and tobacco use over a 16-year period of late adulthood, the investigators learned that continuing smokers also

Drown Your Sorrows

15%	smokers who have a current problem with alcohol
40%	smokers who have a history of problems with alcohol
80%	alcoholics who currently smoke
9%	those in the general population who are heavy smokers
72%	alcoholics who are heavy smokers
15% – 20%	heavy smokers who are alcoholic
21%	those in the general population who are light smokers
11%	alcoholics who are light smokers
28%	those in the general population who are ex-smokers
17%	alcoholics who are ex-smokers
42%	those in the general population who have never smoked
8%	alcoholics who have never smoked
80% – 85%	recovering alcoholics who neither crave alcohol more nor relapse to alcohol use when they quit smoking
23%	alcoholism treatment providers not in favor of encouraging clients to stop smoking
46%	alcoholism treatment providers who never encourage their clients to stop smoking

used increasing amounts of wine. Alcohol consumption did not change across time in nonsmokers.

Carmelli's colleague Gary Swan and his collaborators in 1996 identified a genetic factor apparently underlying the joint use of tobacco, alcohol, and coffee. A report on research with 356 pairs of middle-aged twins, about half of them monozygotic (identical), concluded that the most parsimonious interpretation of the twins' use of tobacco, alcohol, and coffee was a "common genetic latent factor" that would explain the association between use of the substances. A 1997 report by the same team found similar results in an examination of self-report data from nearly 4,600 twin pairs.

Such relationships, however, are not necessarily predictive. A multi-site research team from the universities of Michigan, Minnesota, and Nebraska had hoped that a measure of their participants' alcohol intake would predict the amount of tobacco use approximately four years later. Instead, Faryle Nothwehr and colleagues found that not only did alcohol consumption not predict later tobacco use, but those who quit smoking were no more likely to reduce their alcohol use than were continuing smokers. Therefore, even though alcohol use and smoking are highly related, quitting smoking did not necessarily reduce alcohol consumption. Men in the study drank more alcohol and smoked more cigarettes than their women counterparts. Among nondrinkers, men were more likely than women to quit smoking. Surprisingly, drinkers and non-drinkers used about the same amount of cigarettes. Those who drank heavily smoked more than did lighter drinkers. And, not surprisingly, smokers used more alcohol than did nonsmokers.

The findings were concordant with a 1993 study in which Perkins and associates reported that middle-aged women who smoke or who used to smoke consume half again as much alcohol as do women who have never smoked. A history of using cigarettes also was associated with dietary and activity patterns that heightened the risk of some chronic diseases. Current smokers reported engaging in less physical activity than did never-smokers and ex-smokers. Quitting smoking was associated with more physical activity.

These connections between commonly used substances can be problematic for a smoker who wants to quit using tobacco, or an alcoholic who is attempting to quit drinking. It is commonly but perhaps inaccurately believed that cigarettes provide help for alcoholics dealing with the stress of abstinence from alcohol. However, it may be that because smoking and drinking so often go together, smoking could act as a trigger, prompting a recovering alcoholic to resume drinking.

Quitting drinking is not always associated with quitting smoking; also, the use of either substance involves a known health risk. Richard Hurt and his colleagues at the Mayo Clinic studied a group of alcoholics who stopped drinking but did not stop smoking. A follow-up 20 years later found that almost half had died. Less than one-fifth of the general population normally would have died by that age. More than half of the deaths of the persons in treatment were due to tobacco-related disease. One-third died from alcohol-related disease.

Treatment Issues

Until the last few years, the mental health treatment community had little interest in promoting smoking cessation among patients with serious psychiatric disorders. In the words of researcher Michael Resnick, "Psychiatry is belatedly awakening to the seriousness of nicotine addiction." It is also belatedly realizing that psychiatric patients need, deserve, and can benefit from effective smoking cessation treatment.

The mental health community used to insist that psychiatric patients should not be given the additional challenge of quitting smoking, since it might disrupt treatment. They were perceived as being too fragile to quit, or as lacking sufficient insight to make them good candidates for quitting. In some psychiatric settings, it was considered beneficial for staff and patients to smoke together as a way to bridge communication barriers. Staff used cigarettes as a means of rewarding patients. Patients were expected to tolerate the environmental tobacco smoke in psychiatric units, even if it triggered a relapse in those who were ex-smokers.

When hospitals across the United States went smoke-free, many psychiatric units did not, as Resnick described the events. Typically, the rest of the hospital went smoke-free before the psychiatric unit did. When, in the late 1980s, hospitals began controlling and limiting smoking in psychiatric units, the results were encouraging. Chaos did not erupt. Staff, initially skeptical, became supportive. Some patients were given nicotine replacement treatment. Most patients were compliant. The air cleared. Staff no longer used cigarettes as reinforcement, patients and staff were no longer exposed to smoke, and hospitals could come completely into compliance with accreditation requirements that hospitals be smoke-free.

Providing effective smoking cessation programs for psychiatric patients requires that clinicians be familiar with both psychiatric treatment *and* addiction treatment. Many aspects of cessation treatment with these groups remain unexamined. Among these is the problem of convincing the patient to accept treatment and remain abstinent in a social environment outside the clinic that encourages or tolerates smoking. Psychiatric medications need to be considered and managed carefully, because of interactions with nicotine. No one yet knows which patients will be best suited for quitting on their own, when is the best time to offer cessation treatment, or exactly how to adjust treatment models to the needs of these patients. Despite those unknowns, current successes reported with psychiatric patients are encouraging.

At What Price?

Whatever effects nicotine has on a smoker require continual rejuvenation. Nicotine cannot fix or cure psychiatric symptoms. The euphoriant effect of smoking is not enormous, nor does it last long. It comes and goes within about 10 percent of the length of time that a cigarette lasts, the Pomerleaus estimate. What price do smokers pay for this temporary help, or this small effort toward emotional and mental steadiness?

As a smoker will tell you, it's the coping enhancement that makes tobacco so appealing, especially to people whose life problems and psychiatric symptoms are difficult to manage. For some smokers, tobacco is that nudge, that little indulgence that makes nightmares livable. Life has many such small regulators, however, such as tucking your cold toes under a snoozing dog, even when the rest of you can't get warm. Or petting your cat while you watch a sad movie. Or melting a piece of fine chocolate in your mouth while you mutter at your income tax forms. Independent of nicotine's addictive properties, it can be this sort of emotional stability, however small, however fleeting, that tobacco affords. It is one thing that keeps people smoking who have precious few other ways to cope. For many smokers, nicotine is also a glue that helps hold them together.

The price smokers pay for these gestures toward stability is, unfortunately, high. Their fate, however, is not without historical analog. It was the well intentioned, hapless Uzzah, we read in the Bible (2 Samuel 6:6-7), who drove a cart carrying the Ark of the Covenant from the house of Abinadab. When the ox stumbled at a threshing floor, Uzzah put forth his hand to steady the Ark.

For this, we learn, God struck him dead.

CHAPTER 6

With friends like this . . .

"The thing about smoking is, it's your best friend," she states. "Okay, it's killing you. But it's your best friend.

"It's not a conscious thing. It's always there for you. When you're happy, it'll help you celebrate. When you're sad, it's there to comfort you. It's there to help you study; it's there to help you work; it's there to help you play. It's always there for you. When you quit, you're giving up a part of your life."

It becomes automatic. A part of you. As integrated into your movements as a common gesture—or a limp?

"You're addicted; you don't think about it. You don't think about how to inhale and how to hold a cigarette, and how to smoke, you just do it. It's like eating. You don't think about how to eat, unless you're in a fancy-schmancy restaurant and you need to be more aware of your manners. If you're home, eating, or drinking a glass of water, you don't think about it, you just do it. You've trained your body to do it without thinking about it. It's the same thing with smoking.

"It's your buddy. It helps you think, helps you cope, helps you stay thin. . . .

"With friends like this, who needs enemies, right?"

═══════════

A Lethal Diet

They are the "super-gainers," and becoming one is many smokers' worst nightmare. When most smokers quit, they gain an average of 8 to 10 lb—which is usually unpleasant, but manageable. When super-gainers quit, they pack on 20 or 30 lb. Some of them carry it around the rest of their lives. Even beyond the health risk this extra weight might pose to them, it stands as a warning to other smokers: *Beware all who enter here. You, too, could "balloon out."*

Exactly why the super-gainers add so much weight when they quit is still being explored. What is known is that this excessive weight gain appears to be genetically driven and appears to be the exception rather than the rule. Different studies' estimates of an average smoker's post-cessation weight gain vary by a few pounds, but the reports all fall within the range of approximately 6 to 12 lb.

David Williamson of the U.S. Centers for Disease Control and colleagues from the CDC and the National Center for Health Statistics first described the super-gainer phenomenon in 1991 in *The New England Journal of Medicine*. Analyzing longitudinal data from more than 2,600 men and women, the investigators found that the average weight gain that could be attributed to smoking cessation was about 6.2 lb in men and about 8.4 lb in women. However, a weight gain of more than 28.6 lb occurred in nearly 10 percent of the men and 13 percent of the women who quit smoking.

Those at the highest risk for excessive weight gain were African-Americans under the age of 55 and smokers who consumed 15 or more cigarettes per day. Among women, being underweight and having a sedentary lifestyle were related to major weight gain. The finding that African-Americans who quit smoking tended to gain more weight than persons of other races was unexpected, since race otherwise has little effect on weight gain among nonsmokers, former smokers, or continuing smokers.

Williamson and his colleagues also found that the average body weight of smokers who quit did not exceed that of individuals who had never smoked. Rather, quitting smoking merely brought that group up to the average weight of nonsmokers. In other words, because of nicotine's effects, a smoker weighs several pounds less than if he or she did not smoke. When the smoker quits using nicotine, body weight returns to what would be normal for that person.

The researchers emphasized that the average person who quits smoking will gain less than 10 pounds, and that about half of the quitters will gain even less. "The beneficial effects of quitting smoking are not likely to be negated by the weight gain that may follow," they wrote.

In a companion commentary to the Williamson article, researcher Neil E. Grunberg of the Uniformed Services University of the Health Sciences raised a possibility that he and others were exploring: "Some investigators believe that the reason nicotine and other drugs of abuse are addictive is that these drugs affect mechanisms controlling body weight and appetite and thereby come to be interpreted as food." Grunberg added that postcessation weight gain "does not mean that potential weight gain is a sound excuse to continue smoking, because the health benefits of smoking cessation exceed the risks of weight gain associated with quitting."

Exactly what causes this relatively uncommon response to quitting smoking remains unknown. Researchers Gary Swan and Dorit Carmelli of SRI International determined that the phenomenon may be influenced by underlying genetic factors. By studying smoking and weight changes in more than 2,100 men in the National Academy of Sciences-National Research Council Twin Registry, in 1995 they confirmed the existence of the super-gainers previously identified by the Williamson group.

They examined the likelihood of being a super-gainer in twin pairs and found that identical (monozygotic) twins were more likely than nonidentical twins to both be super-gainers. However, they also noted several additional relevant factors, including marital status and alcohol consumption. Super-gainers were more likely to be unmarried, and thus may have experienced less outside influence in managing caloric intake and being weight conscious. Also, super-gainers' liquor consumption increased more than twice as much as did that of the average ex-smoker.

"The large increase in consumption in super-gainers suggests the possibility of an especially strong compensatory mechanism," Swan and Carmelli wrote, "with alcohol playing a central role in replacing the effects of nicotine at either the neural-genetic or the behavioral level."

The researchers also commented on the existence of a subgroup of former smokers who reported losing—yes, losing—weight after quitting smoking. These smokers were older when they quit, they had a higher body mass index (proportion of body fat) than other smokers, and they were more likely to be retired and to have cardiovascular disease. Swan and Carmelli proposed that this group's weight loss could have resulted from aging, from disease, or from efforts to reduce the progression of disease by losing weight.

The Weight-Control Smoker

Weight gain after quitting apparently results both from metabolic adjustments the body makes and from increased consumption of food and alcohol after quitting smoking. Even if smokers do not take in more calories, they probably still will gain some weight. Using nicotine replacement during smoking cessation delays the weight gain, although the eventual gain will be about the same as if the smoker had never used the alternative source of nicotine. Even so, this delay may provide some benefit, since it allows the smoker to deal with one problem at a time. First, he or she can focus on quitting smoking and staying quit. Then, some weeks or months later, the ex-smoker may be better able to manage the weight gain that might accompany tapering off the nicotine replacement.

Using tobacco as a weight-control mechanism is not a universal motivation, but it is a common one. Nicotine suppresses appetite and alters metabolism in such a way that someone taking up smoking is likely to drop some weight. Smoking, particularly at a high level, also is believed to delay the rate at which solid foods empty from the stomach, an effect identified by Ellen Gritz and others working with researcher Murray Jarvik at the University of California at Los Angeles. Many smokers, primarily women, use nicotine as a way to control weight, as researcher Cynthia S. Pomerleau and colleagues of the University of Michigan found when they profiled the "female weight-control smoker."

Several scientific reports have shown that although weight-control smoking is rare in male smokers, it occurs in as many as 40 percent of female smokers. Identifying smokers with this profile highlighted the cultural impact on concerns about weight, since (in Pomerleau's words), "[A]fter all, no rat would be likely to self-administer nicotine to maintain a svelte profile, nor would individuals living in a culture that valued plumpness (e.g., as a symbol of prosperity) be likely to respond to this property of nicotine."

According to Pomerleau, it appeared that the tendency of many girls and women to smoke as a means of controlling weight had several bases. First, slenderness is valued throughout much of Western society. Second, women appear to be more sensitive than men to the effects of nicotine on their food intake and their weight. Additionally, women have a higher expectation that nicotine will help them control their appetite and weight.

The researchers examined whether there is a subgroup of smokers,

"A Moment on the Lips . . ."

4 – 7 lb less	what a smoker weighs, compared with a never-smoker
same	what an ex-smoker weighs, compared with a never-smoker
less than 10 lb	how much weight the average smoker gains after quitting smoking
10%	quitters who might gain as much as 30 lb after quitting
negligible	health risk associated with typical weight gain after quitting smoking
75 – 100 lb	how much weight a smoker would have to gain to achieve a health risk equivalent to smoking one pack a day

primarily women, who would be likely to be "weight-control smokers." With items embedded in a questionnaire, they identified women who endorsed statements indicating that they smoked to keep from gaining weight or to control their appetite. Women who scored high on the weight-related smoking questions were compared with women who did not report that smoking helped control their appetite or that they smoked to control their weight. Women who were smoking to control their weight reported that when they had been abstinent from tobacco in previous quit attempts, their appetite and weight had changed.

As Pomerleau was quick to note, the ways in which the weight-control smokers and their counterparts did not differ were as interesting as the ways in which they did. The investigators found no evidence that the weight-control smokers were more depressed or anxious. The weight-control smokers were no more nicotine dependent than were those who did not smoke to control weight. However, they had a greater tendency to use nicotine to trigger changes in their internal state, and perhaps an enhanced tendency for such changes to occur.

The researchers also measured withdrawal symptoms in a subsample of the same women. The team wondered whether the weight-control smokers would experience abstinence from tobacco differently than other women smokers would. Of the nine withdrawal symptoms measured in the groups of women over two days of abstinence, only "increased eating" was greater in the weight-control smokers. The researchers inferred that the weight-control smokers might have patterns of excessive or unpredictable eating, which nicotine might help them control.

Pomerleau and her colleagues recommended several possible changes in smoking cessation strategies, based on their findings and those of other scientists. They proposed that researchers should consider studying the following approaches to cessation with these smokers:

- Targeting weight-control smokers so that more attention can be given to their needs in cessation programs.
- Prescription of nonnicotine medications that can prevent weight gain, either alone or with nicotine replacement that is slowly discontinued through tapering.
- Therapeutic behavioral techniques that help in dealing with "disinhibited" or binge eating. These could be taught in a formal stop-smoking program.

Gains and Losses

Scientists also have considered whether concern about weight gain would affect the likelihood of success in smoking cessation. Andrew W. Meyers and his associates at the University of Memphis and the University of Alabama at Birmingham found that persons who used formal smoking cessation programs as a way to quit were less concerned about weight than the average smoker. Also, those who were concerned about weight and who sought help to quit smoking were less likely to quit smoking. These results confirmed findings by Meyers' colleague and co-author Robert C. Klesges of the University of Miami, who previously had found that many smokers were concerned about weight gain, but those who were the most concerned had the least intent to quit smoking. Klesges and his co-investigators found concern about weight gain in both men and women smokers, and observed that those who anticipated the gain were more likely to relapse.

Although the connection between tobacco use and weight control has been recognized for at least a century, it is only in the last decade or so that researchers have been able to study the association in detail. Some common assumptions have not yet been demonstrated, including these: (1) Smokers keep smoking to prevent a weight gain that might result from cessation. (2) Weight gain is a cause of smoking relapse. (3) Controlling weight gain during cessation prevents relapse.

Actually, the opposite of those notions may reflect the reality of smoking cessation. For example, stop-smoking interventions also designed to help prevent weight gain have even been found to heighten the risk of relapse. The current thinking is that as long as weight gain does not cause relapse, weight gain can be managed more easily if it is handled after smoking cessation is no longer the most pressing issue.

The most recent findings indicate that smoking cessation results in weight gain primarily through increased caloric intake after quitting smoking. During the first month after a smoker quits, food intake increases by some 300 to 400 calories per day, with much of this increase due to snacks between meals. This increase accounts for the approximate pound per week that the average ex-smoker gains after quitting. These effects may be particularly noticeable for women, who tend to obtain more postcessation calories from snacks than men do, and who have been found to increase food intake more than men do after quitting. Women smokers who are high in dietary restraint (in other words, women who have chronic concerns about weight and who diet to maintain an unreasonably low body weight) may use smoking as a way to suppress eating; for them, weight control is perceived as a benefit they derive from smoking.

The phenomenon of using smoking to suppress weight emerged among both men and women smokers in a 1996 study of smokers in Austria. Éva Rásky of the Institute of Social Medicine, Karl Franzens Universität at Graz, Austria, and co-authors found in a study of more than 27,000 rural Austrians that light or moderate smoking was correlated with a lower weight (relative to height and build), while heavy smoking and quitting smoking were related to higher weight.

In U.S. adults, smoking cessation does contribute to the prevalence of overweight, although the impact is small. Katherine M. Flegal and colleagues of the U.S. CDC reported in 1995 that weight gain associated with quitting smoking accounted for about one-fourth of the weight increase noted in men and about one-sixth of the increase seen in women between 1978 and 1990 in the United States. Taking into consideration other factors such as age, demographics, physical activity, alcohol use, and childbearing, they found that men in their national survey gained an average of 9.8 lb, and women gained an average of 11 lb. Smokers who had quit during the previous decade were more likely to become overweight than were nonsmokers.

Since weight control is important to many smokers, offering weight-control help in conjunction with a cessation program seems appropriate. However, an extensive clinical trial by Sharon Hall and her colleagues at the San Francisco Veterans Affairs Medical Center and the University of California at San Francisco, reported in 1992 that the opposite may be the case. The research team offered three types of adjunct treatment in conjunction with stop-smoking treatment: (1) a behavioral weight-control program that included exercise, weight monitoring, and calorie control; (2) a nonspecific weight-control group involving group therapy, with supportive help and information on nutrition and exercise; and (3) information packets on nutrition and exercise. Fewer people remained abstinent in the two weight-control groups than in the third group at three months, one year, and longer than one year.

A second reason for weight gain following cessation is the change in metabolic rate that accompanies the use of each cigarette. Nicotine does increase the metabolic rate (or *resting energy expenditure*) of tobacco users. Janet Audrain and colleagues reported in 1995 that although nicotine heightened the metabolism of both normal-weight and overweight women, nicotine's effects were attenuated in those who were overweight. Thus, overweight women might not be obtaining as much weight-control "benefit" from nicotine as they believe they are receiving.

This report concurred with findings from a research team at UCSF, who noted that although smokers weigh less than average, smokers who smoke more weigh more than those who smoke less. The investigators, Lidia Arcavi and colleagues, found that nicotine increased heart rate and energy utilization in most smokers, but that these effects were most pronounced in smokers who used 10 or fewer cigarettes a day. These low-level smokers also used more energy as a result of their nicotine use than did high-level smokers consuming 15 to 30 cigarettes per day.

Curiously, the researchers also found that the phenomenon of tolerance was different among low- and high-level smokers. (As explained in chapter 3, developing tolerance for a substance involves needing increasingly larger quantities of it to achieve the same drug-related effects.) The low-level smokers developed tolerance as indicated by acceleration in heart rate and in energy expenditure. High-level smokers, however, developed only tolerance indicated by heart rate. The tolerance to nicotine seen in their cardiovascular response was not matched by changes in energy expenditure; either they had a rapid development of tolerance or they showed no effect at all. The authors concluded that these differences between groups of smokers could help explain what they termed the "unusual" relationship between nicotine and body weight.

Influences on Women

Overall, nicotine dependence may be a somewhat different experience for women than for men. The appetite-suppressant capacity of nicotine may increase its appeal to women and thus increase women's possibility for developing dependence. Additionally, significant evidence indicates that individuals prone to depression are also more likely to use nicotine. Since depression is more common in women than in men, this becomes an issue of particular concern for women smokers. Many aspects of the nicotine-weight relationship seem particularly problematic for women.

A 1994 article by John Pierce and associates at the University of California at San Diego Cancer Center explored trends in smoking initiation among children and adolescents, focusing in particular on the effects of targeted advertising. In their report, which was part of an issue of *The Journal of the American Medical Association* devoted primarily to tobacco research, they noted that the increase in smoking prevalence among girls younger than the legal age for buying tobacco "started the same year that the tobacco industry introduced women's brands of cigarettes." That year was 1967. The researchers' examination of the temporal (time-related) correspondence between advertising campaigns and smoking initiation among women showed that the increase was notably higher in women who had never attended college. Across the period under study, 1944 through 1988, the rates at which adolescent boys started smoking changed little. The authors concluded that the increase in smoking among young women was associated with increased advertising targeting women.

Another risk factor enhanced in women smokers is the prevalence of depression. Women are more prone than men to experience depression; they also are at heightened risk of becoming nicotine dependent. This does not mean that women are biologically predisposed to becoming more physically dependent on nicotine, or that dependency is more a basic part of their nature. Rather, it means that some of the predisposing factors are more common to women.

Despite women's increased risk for problems that might contribute to dependence, smoking rates remain higher among men than among women almost universally throughout the world. An exception is the Lahanan people of central Borneo, a group of about 300 who live in a horticulture-based economy that has given them only limited contact with the world outside their community. In this setting, the Lahanan women traditionally have been the primary cultivators and disseminators of tobacco, and they also have been the heaviest users.

Hand Me the Ashtray, Dear

24 – 20   ratio of male smokers to female smokers in the United States

70%   men who smoked in the 1940s and 1950s in the United States

50%   men who smoked in the 1960s in the United States

28%   men who smoke currently in the United States

18%   women who smoked in 1935 in the United States

34%   women who smoked in 1965 in the United States

23%   women who smoke currently in the United States

32%   men age 25-44 who smoke in the United States

13%   men 65 and older who smoke in the United States

28%   women age 25-44 who smoke in the United States

11%   women 65 and older who smoke

1986   year in which lung cancer surpassed breast cancer as the leading cause of cancer death among U.S. women

2000   year by which U.S. female smokers are projected to outnumber male smokers

This reversal of the usual sex-related tobacco-use pattern is changing, however, as the group has more contact with the industrialized world. Smoking among these women is becoming less prevalent, and men are taking up cigarette smoking rather than using the home-grown tobacco. Young persons who acquire an education tend to quit using tobacco, according to researchers who have studied this group, Jennifer and Paul Alexander of the University of Sydney in New South Wales, Australia.

Aside from economic and cultural considerations, women respond somewhat differently to nicotine than men do. Ken Perkins and colleagues of the University of Pittsburgh found in 1994 that women and men smokers reported nicotine's "subjective" effects differently, although both men and women were able to tell the difference between a dose of nicotine and a dose of non-nicotine placebo. Women reported dose-related nicotine effects in feelings described as "dizzy," "stimulated," "jittery," and "head rush," in response to increased doses of nicotine, but men did not. Nicotine was administered in relation to each subject's weight, via an inhaler device. The researchers used substances to mask the taste and smell of nicotine and the irritation of nasally administered nicotine.

Sex-based differences in nicotine's effects also can be related to the situations in which nicotine is used. Perkins noted in a 1996 review article that women may be more sensitive than men to situational factors such as the sight and taste of cigarette smoke. Research has shown that men and women may smoke for different reasons. Men apparently smoke to maintain a steady level of nicotine in their body, but women might be smoking to obtain effects that are less related to nicotine.

Tobacco is known to contain thousands of chemical compounds and is associated with many complex factors that can elicit their own responses in addition to those elicited by nicotine. In other work, Perkins' team noted that subjects responded differently to smoked tobacco and nasally inhaled nicotine spray, which suggested that smoking generates effects in addition to the delivery of nicotine. Even though most of the responses appear to be due to the delivery of nicotine, other aspects of smoked tobacco cannot be discounted or ignored.

Such gender-related differences in what makes smoking reinforcing could influence the effectiveness of a stop-smoking program in helping women quit. It is possible that women might not find nicotine replacement (such as gum or patch) as useful as men would find it, and that cessation geared generically toward both sexes could overlook women's particular needs in smoking cessation.

Males and females also appear to differ in the way nicotine affects the body's energy balance. It is possible that smoking helps some women suppress appetite and eating by adding the reinforcing effect of an increase in energy. Nicotine increases the body's expenditure of energy, or its metabolism. When male smokers engage in light physical activity, metabolism is enhanced. Women smokers, however, experience no metabolic enhancement from light activity. This gender difference, as explained by Perkins in a 1997 chapter, existed only during physical activity and was not present when subjects were at rest.

Each time a smoker used tobacco, the nicotine prompted a brief boost in metabolism of 5 to 7 percent for about a half hour. A pack-a-day smoker was likely to boost at-rest metabolism throughout the day, even though the effects were brief, because he or she used nicotine so continuously. A smoker using caffeine boosted metabolism even more, since the effects of the two substances (caffeine and nicotine) were additive. Men who engaged in light activity while using caffeine and nicotine together experienced an enhanced metabolic rate. Women did not.

Who is most likely to have a metabolic boost from smoking? The answer: men in good physical condition who typically engage in high levels of activity. Women, who are most likely to use nicotine to lose or maintain weight, are less likely to experience a metabolic benefit from using tobacco.

The use of nicotine replacement in cessation can forestall weight gain. Also, exercise and dietary changes can influence success in quitting smoking and can help a smoker avoid gaining an undue amount of weight. Even so, quitting smoking and dieting at the same time might not be the best approach. Several studies have found that combining smoking cessation with weight-control methods can worsen the abstinence rate, canceling out any presumed advantage of tackling both issues at once. Dieting can result in some of the same noxious effects as smoking abstinence, such as mood disruption, diminished arousal, and fatigue.

Another male-female difference that can affect smoking rates and success in quitting is the effect of nicotine in reducing feelings of stress. Women report a greater tendency than men to smoke as a way to reduce negative emotions. This could also make women more prone to relapse during stressful times. When a woman is attempting to quit smoking, she may find that smoking relieves her feelings of being stressed. It is possible, however, that the stressful feelings are due primarily to abstinence symptoms (or withdrawal), and thus the stress relief that tobacco provides during cessation could be due to nothing more than relief from those symptoms.

All of these factors could be part of the overall statistical picture reflecting these facts:

- Smoking has not declined as much in women as in men.
- Young women are taking up smoking at a greater rate than are young men.
- Quitting may be more difficult for women than for men.
- Women may be more likely to relapse.

All Is Not Lost

These findings about weight and smoking, particularly as they affect women smokers, can be discouraging. Even so, smoking—present, past, or passive—is only one of many factors affecting weight. Swiss researchers Martine Bernstein and colleagues of the University Canton Hospital at Geneva, Switzerland, examined the relationship between education, smoking status, and weight in 928 Swiss women. They found that a woman's education level was an important predictor of her current weight and her weight history (i.e., previous weight and weight gain since age 20). Smoking status appeared to have little effect on weight. This finding is ironic in light of findings that women are more likely than men to report using tobacco as a way to control weight.

Bernstein and her colleagues found that the group of women (with ages ranging from 29 through 74) who had the most education weighed the least. The least educated group weighed an average of 9 lb more than the most educated group. Differences in relation to tobacco exposure status were small. Women exposed to secondhand smoke, whom the researchers termed "passive smokers," weighed the most at 140 lb, and former smokers weighed the least at 133 lb. Weight differences associated with education status were greater.

The research team concluded that although smoking "may influence short-term weight variation," it had little long-term effect on weight. They explained: "The finding that smoking is not an efficient means of weight control has major implications with respect to public health strategies aimed at reducing smoking. Women will be less reluctant to quit smoking if they know it will not promote a long-term weight gain."

Several strategies could be helpful for a smoker concerned about weight and also contemplating quitting. Perkins and his colleagues in 1997 proposed what they termed a "cognitive-behavioral" approach to cessation that would treat not the weight gain itself, but the smoker's concerns about weight gain. As the term itself implies, *cognitive-behavioral* refers to a treatment that helps people change both their thinking and their behavior. In this case, a goal would be for participants to change their beliefs about the primacy of not gaining weight, and clinicians would work with them to change the behaviors associated with smoking, to facilitate quitting.

Perkins explained that concern about the typical weight gain of 8 to 10 pounds "must be viewed as dysfunctional and unreasonable in light of the health risks of continuing to smoke." A cognitive-behavioral approach would assume that the attitudes and perceptions of the importance of weight need to be modified, not the tendency to gain weight. "Weight concerns may stem from unreasonable perceptions about the weight gain and, more importantly, from the heightened importance placed on a modest weight gain in controlling one's health behaviors (i.e., whether or not to quit smoking)," Perkins added.

Other options include pharmacological (that is, drug) treatments. In this regard, smokers have several options. First, smokers who quit with the help of nicotine replacement are likely to find that substituting one source of nicotine for another delays the weight gain often associated with quitting. Once they taper off the nicotine, they may find that they then gain some weight. However, by this time they may be better prepared to deal with the weight gain than if they had attempted to quit smoking and severely restricted their diet at the same time.

Studies also have examined the usefulness of a drug called phenylpropanolamine, which is sold without prescription in the United States both as a decongestant and as an appetite suppressant. To date, phenylpropanolamine has been examined only in short-term use, so its effects on long-term abstinence and long-term weight gain are undetermined. Typically, drugs that suppress appetite do not continue to prevent weight gain once their use is discontinued.

A relatively simple but effective strategy involves changes in levels of exercise. A moderate increase in physical activity level minimizes postcessation weight gain without requiring extreme restrictions on food intake. Several studies indicate that it is not necessary to undergo strenuous exercise to achieve enough metabolic enhancement to keep postcessation weight gain within manageable levels. The addition of such common activities as walking instead of driving short distances, or climbing a flight or two of stairs instead of taking an elevator, can help keep the weight gain to levels below the usual Thanksgiving/Christmas increase most celebrants experience every holiday season.

The worst strategy, it appears, is to attempt to control postcessation weight gain through undue restrictions on food intake. One of the more interesting findings leading researchers to this conclusion was the discovery by Hall and her colleagues in 1986 that smokers who gained the least weight were the most likely to relapse back to smoking.

Evidently, the relationship between smoking and body weight is not simple; what is simple, however, is the fact that the benefits of quitting smoking outweigh the disadvantages of the weight gain that might accompany quitting.

CHAPTER 7

She calls it the "grand gesture of smoking."

"Smokers do it unconsciously," she explains, throwing her head back and exhaling upward with a flourish. "It's a very unconscious movement, dragging it, and then blowing it up, and throwing your head back." She mimics a silent-movie star, blowing imaginary smoke into the air.

"People who don't smoke make fun of the glamour. To a real smoker, there is no glamour. It's just a thing. It's like eating and drinking water. It's just there. It's time to go have a cigarette."

She shrugs and picks up a cigarette and lighter. It's time to go have a cigarette.

"I try to be a considerate smoker," she mentions as we walk outside. "I don't like to smoke in my house, and I don't allow smoking in my house. I don't like the furniture, and the drapes, and the carpet and the clothes smelling like smoke. The only smoking at my house is on the front porch and the back porch."

Sometimes, to their condemnation, some people smoke on her back porch with the back door open, not noticing which way the wind is blowing. If it blows into the house and she can smell it at the other end of the house, she gets irritated, she says.

As the city winds shift, she moves from one side of me to the other. She is a downwind smoker.

"I try to keep the smoke away from the people who don't smoke. Like . . . the wind is blowing this way right now, so, technically I should be on the other side of you." She moves around me again.

"I try to be really aware of when somebody's walking by and the wind is blowing toward them, so that I don't take a drag on my cigarette until they're past.

"I think it's really funny when I'm standing there, just holding a cigarette, and the winds are blowing from west to east, and somebody who's walking from east to west passes me, so they're upwind of me, and I have a drag, and they go [cough] [cough] like that. They can't even smell it, so what is their problem? The wind is, like, 20 miles an hour the other way. I find things like that amusing."

She particularly doesn't like it when other smokers blow smoke in her face.

The Accidental Smoker

The nonsmoking section was full by the time the woman got a seat on the plane. In the 1970s, before U.S. airlines went smoke-free on domestic flights, it was still possible for a non-smoker to be stuck in the middle of a smoking section. She resigned herself to what she knew would be several hours of discomfort and buckled herself into her seat.

As she noticed cigarette packs bulging the shirt pockets of the men seated around her, she thought back to her childhood, when her father's boss would occasionally come over to spend the day at their house. He would chain-smoke, filling their small home with an odor that her mother would spend days trying to air out of the draperies and carpets. They had never dared ask him not to smoke, because he was the boss.

And she thought back to the job she had left two years earlier, where she had worked only a few feet away from a chain-smoker, eight hours a day. She had developed asthma and, on her doctor's advice, had quit that job to take another. Ever since then, the smell of cigarette smoke had made her queasy. It also made her cough. She would rather have been sentenced to cleaning latrines than to spending the afternoon in the smoking section of an airplane. She mentally chided herself for not reserving a seat in the nonsmoking section.

Once the flight was airborne, the man next to her pulled out his pack of cigarettes and tapped one free.

"Do you mind if I smoke?" he asked, almost as an afterthought.

"Not if you don't mind if I get sick," she answered.

This small encounter, now anachronistic in many social settings, typifies the war that erupts repeatedly over involuntary exposure to tobacco smoke. Smokers claim a right to use their tobacco, sometimes without seeming to be aware of just how noxious others might find their smoking to be. Those on the other side want to avoid exposure to environmental tobacco smoke at almost any cost. Compromise can be difficult to achieve.

The facts, as we now know them, are these:

- Exposure to secondhand or environmental tobacco smoke can present serious hazards to the health of pets, children, and adults. Evidence supporting this belief is not without controversy, but at this time, the fact that environmental smoke poses health risks is generally accepted in the scientific community.
- Exposure to such smoke also is a source of considerable discomfort for many persons, independent of any health risks. It can result in nausea, coughing, watery eyes, smelly hair, stinky clothing, and stale-smelling household furnishings. Many nonsmokers do not like to be in settings where smoking is allowed. If a nonsmoking friend comes to a social gathering where smoking is allowed, the host should consider the nonsmoker's presence to be a great compliment.
- When it became socially acceptable and legally feasible to object formally to tobacco smoke, many groups who had previously been quiet about secondhand exposure became vocal. These included airline flight attendants, waiters and waitresses, hospital

employees, passengers on mass transit, and, in some cases, spouses of smoking partners. A lot of nonsmokers, as it turned out, have not liked tobacco smoke for a long time.

What It Is and Does

The terminology associated with this phenomenon can be confusing. The most descriptive overall term is *environmental tobacco smoke* (or ETS), which includes *sidestream smoke* that comes from the burning cigarette (for example, between puffs) and *mainstream smoke* that is exhaled by a smoker. The term *passive smoking* refers to inhalation of tobacco smoke in the air, and is synonymous with *secondhand smoking*. Most environmental tobacco smoke consists of sidestream smoke. Although mainstream and sidestream smoke differ, "active and passive smokers inhale the same toxins and are thus likely to suffer from the same health effects," according to German tobacco expert Friedrich J. Wiebel. He listed the immediate effects of environmental tobacco smoke as the following, in a 1997 review:

- Sore eyes
- Itching
- Sneezing
- Runny nose or stuffiness
- Sore throat, cough, wheezing, hoarseness
- Upset stomach (can last as long as 24 hours)
- Dizziness (also can last as long as 24 hours)
- Headaches (also can last as long as 24 hours)

The chronic effects can include the following:

- Chronic respiratory symptoms in children (bronchitis and pneumonia; middle-ear fluid; cough, phlegm, and wheezing; decrease in lung function; asthma)
- Chronic respiratory symptoms in adults (phlegm, cough, difficulty breathing on exertion, bronchitis)

- Fetal toxicity from both passive and active maternal exposure to smoke (nicotine exposure similar to that of a light to moderate smoker; spontaneous abortion; children's asthma; later development of decreased performance in language, speech, and visual-spatial abilities)
- Sudden infant death
- Cardiovascular disease
- Lung cancer

Studies reporting relationships between these diseases, particularly lung cancer, and passive exposure to tobacco smoke have not gone without criticism. As with any area of scientific inquiry, the perfect study remains yet to be done; some epidemiological studies, in particular, are vulnerable to criticism because of inadequate methods and analyses. Despite these acknowledged limitations, Wiebel concluded, "In the final analysis, it is the total weight of the toxicological and epidemiological evidence which gives confidence to the conclusion that ETS is a human lung carcinogen."

Hazards

The use of tobacco entails considerable health risk. Many volumes have been written to document this assertion; in fact, no book can be current about the subject, because new announcements of tobacco's health risks are published continually in medical journals. The risks, detailed in literally thousands of scientific articles covering a broad range of disciplines, have been enumerated for decades and continue to be explored. The general public is probably best acquainted with the risk of cancer, since the earliest public declarations about tobacco's effects focused on lung cancer. However, tobacco-related cardiovascular disease is an even greater cause of death and disease than tobacco-related cancer. The list of health problems that are caused or exacerbated by tobacco is lengthy; it includes the following:

- Coronary heart disease
- Other heart disease
- Stroke
- Other circulatory disease
- Chronic obstructive lung disease
- Other respiratory disease
- Cancer (lung, lip, oral, mouth, pharynx, esophagus, pancreas, larynx, kidney, cervix, uterus, urinary organs, stomach)

The health hazards of involuntary exposure began to be documented years later than those of active exposure. Even so, more than 12 years ago the U.S. government enumerated problems with environmental exposure to tobacco smoke in its 1986 Surgeon General's report, *The Health Consequences of Involuntary Smoking*. Since the 1986 publication of that report, the U.S. government has continued to endorse reports detailing the hazards of environmental exposure. A 1992 report, *Respiratory Health Effects of Passive Smoking: Lung Cancer and Other Disorders*, from the U.S. Environmental Protection Agency (EPA) continued the theme of the earlier report. It explained that determinations of the potential associations between environmental tobacco smoke and lung cancer use a "weight-of-evidence" analytic approach, in accordance with U.S. standards for assessing carcinogen risk, or, in other words, the risk that a substance poses for causing cancer. Criticism of these findings found its way into the scientific press. However, in August 1998 the Associated Press reported that thirteen scientists had been paid a total of more than $156,000 to submit letters and articles critical of such reports to scientific journals, typically without disclosure of the payments. In fact, it added, the total public health impact from environmental tobacco smoke could be greater than the report indicated.

In 1997, the California Environmental Protection Agency released what *Business Week* magazine called "the most devastating report yet" on the dangers of environmental tobacco smoke. The California report indicated that environmental exposure caused as many as 62,000 deaths from heart disease, 2,700 cases of sudden infant death, and 2,600 new cases of asthma each year. It declared that secondhand smoke increased the risk of cervical cancer and spontaneous abortions. It identified more than 50 tobacco compounds as carcinogens, with 6 also listed as sources of reproductive or developmental problems. *Business Week* writers Paul Raeburn and Gail DeGeorge commented, "[T]he powerful evidence in the new reports damning secondhand smoke suggests that, outside the courtroom, there is nothing left to debate."

New findings about serious health consequences of both voluntary and involuntary exposure are published so continuously that it is difficult to stay current about them. A comprehensive 1997 book on the health consequences of smoking (*Cigarettes: What the Warning Label Doesn't Tell You*) had no sooner been published than scientists announced several newly documented hazards associated with smoking. For example, a September 1997 report in the *Journal of the National Cancer Institute* stated that even those who had quit smoking were at increased risk of some cancers as much as 30 years later. Researchers Marilie D. Gammon and colleagues of the Columbia School of Public Health in New York found that as many as 40 percent of cancers of the esophagus and stomach could be tied to cigarette smoking. Their study of more than 1,200 subjects, 554 of them cancer patients, indicated that both current and former smokers more than doubled their risk of adenocarcinoma, a type of esophageal or stomach cancer. Former smokers' risk was decreased only if they had stopped smoking more than 30 years before. The authors stated that the recent increase in adenocarcinoma among older persons could be due to the increase in cigarette smoking earlier in the century.

Scientists and physicians feared that the results might discourage smokers from quitting, since they might assume that the damage had already been done. They emphasized the necessity of continuing to reduce risk by quitting smoking. Also, since tobacco was not responsible for all such cancers, other elements such as diet, medications, and other medical conditions also present an established risk.

Similar reports emerged in the medical research literature relating not only to the risks of smoking itself, but to involuntary exposure. A sobering announcement came in January 1998, when biostatistician

Forearmed?

two-thirds	those in U.S. in 1991 who accepted cigarette-pack warning labels as legal protection for tobacco companies
51%	those in U.S. in 1996 who accepted warning labels as legal protection for tobacco companies
62%	current U.S. smokers who accept warning labels as legal protection for tobacco companies
52%	former U.S. smokers who accept warning labels as legal protection for tobacco companies
46%	U.S. never-smokers who accept warning labels as legal protection for tobacco companies
62%	those under age 30 in U.S. who accept warning labels as legal protection for tobacco companies
43%	those age 65 and older in U.S. who accept warning labels as legal protection for tobacco companies

(Klein, 1997.)

George Howard and associates at the University of North Carolina at Chapel Hill reported that not only did smoking result in heightened risk of atherosclerosis among smokers, but the risk was also increased significantly for nonsmokers exposed to environmental smoke. The investigators, whose findings were published in *The Journal of the American Medical Association*, wrote: "These data represent the first report, to our knowledge, from a large population-based study of the impact of active smoking and exposure to environmental tobacco smoke on the progression of atherosclerosis." They noted that active smoking played "a major role in the progression of atherosclerosis, as did the duration of smoking measured by pack-years of exposure." They added, "The impact of exposure to environmental tobacco smoke on atherosclerosis . . . was also surprisingly large, increasing the progression rate by 11 percent above those not so exposed."

This report came about six months after the American Heart Association's journal *Circulation* published an article indicating that regular exposure to environmental smoke almost doubled the risk of heart disease. A Harvard University research team led by Ichiro Kawachi found in a 10-year study of 32,000 nurses that those with regular exposure to environmental smoke had a 91 percent greater chance of developing heart disease.

Also published in 1997 was a report from the Australian federal government aimed at reducing the risks from passive smoking. Australia's National Health and Medical Research Council had urged governments and employers to reduce the health risks caused by exposure to environmental tobacco smoke, citing "compelling evidence" that such smoke is hazardous. The NHMRC recited these conclusions:

- Passive smoking contributes to asthma in 46,500 Australian children every year; this effect is marked in children of mothers who smoke more than 10 cigarettes per day.
- Children younger than 18 months old face a 60 percent increase in the risk of respiratory illness if they are exposed to environmental smoke.
- Nonsmokers living with smokers have a 30 percent increase in the risk of lung cancer.
- Nonsmokers living with smokers have a 24 percent increase in the risk of heart attack or heart disease.

The report followed a legal struggle between the tobacco industry and the NHMRC the previous year, in which legal technicalities forced the council to drop formal recommendations for new health regulations. Instead, the council issued the report, which reviewed more than 400 studies from around the world.

Give the Kid a Break

12 – 19 weeks   period of gestation in which nicotine binding sites increase in the human fetal brain, a process altered by contact with nicotine

30 minutes   time at which maximum reduction in fetal "breathing" occurs, following mother's use of nicotine

90 minutes   point at which fetal breathing recovers to normal rates

four times   increased likelihood that the son of a mother who smokes during pregnancy will become delinquent

20%   pregnant women who continue to smoke (U.S.)

25%   pregnant women who succeed at quitting smoking (U.S.)

12%   pregnant women who quit during the second trimester (U.S.)

almost none   decline in smoking prevalence of women of reproductive age in 33 U.S. states, 1990-1993

probably none   number of U.S. states likely to meet the "Year 2000" goal of 12% smoking among reproductive-age women

Challenges and Responses

Scientific findings about the risks of environmental tobacco smoke were questioned in a 1995 report from the Congressional Research Service, a division of the Library of Congress. Economists C. Stephen Redhead and Richard E. Rowberg concluded that the statistics from environmental tobacco studies did not support the conclusion that passive smoking involved substantial health effects. They stated, "[E]ven when overall risk is considered, it is very small risk and is not statistically significant." They determined that nonsmokers exposed to low levels of smoke had little or no additional relative risk of lung cancer because of their exposure. They charged that epidemiological studies were plagued with "misclassification and recall bias."

Opposition to the Congressional Research Service report was quick to arise. The debates, like many others of recent years, emerged on the Internet, where World Wide Web sites from both sides took each other to task. A group called California Smoke Free Cities, funded through the state Department of Health Services' allocation of tobacco tax money, explained that the Congressional Research Service determination was made with a "threshold" approach to cancer risk. The threshold theory holds that below a certain threshold of exposure, there is no risk of lung cancer. Public health authorities have not accepted this theory, but it has been employed by the tobacco industry. With this theory, a researcher estimates fewer potential cases of lung cancer deaths in relation to known risks than would be possible by using more traditional means.

The goal of the Congressional Research Service report, however, apparently was not to take sides either with the tobacco industry or with those in the medical community who had declared environmental tobacco smoke to be hazardous. The report was not in direct competition with the 1992 EPA report, which, unlike the Congressional Research Service report, underwent extensive scrutiny in peer review from independent scientists. The EPA report, unlike the Congressional Research Service report, also received considerable support from federal

Innocent Bystanders

3,000–5,000	U.S. people dying anually from lung cancer associated with involuntary exposure to tobacco (secondhand smoke)
30,000–50,000	U.S. people dying anually from heart disease associated with involuntary exposure to tobacco
double	extent to which secondhand smoke increases the risk of heart disease among women
25%	U.S. children's bed confinement days attributable to secondhand smoke
25%	U.S. children's school absence days attributable to secondhand smoke
24%	nonsmoking parents of an asthmatic child failing to attend an asthma education program in one Minnesota community
42%	parents in one-smoker households failing to attend the same program
78%	parents in two-or-more-smoker households failing to attend the same program

agencies and health-related organizations. Actually, the Congressional Research Service report might have gone largely unnoticed except that the National Smokers Alliance, an organization related to tobacco interests, used the report in an effort to amend a California law requiring smoke-free work environments.

The Congressional Research Service report was not the first challenge to the official U.S. government stance that the public health community's concerns about environmental smoke are warranted. A 1995 exchange in the journal *Risk Analysis* also chronicled the arguments on both sides. In that debate, Gio B. Gori took issue with the 1992 EPA report's conclusions that environmental tobacco smoke causes cancer and increases the risk of respiratory disorders. At issue was a question of whether environmental smoke differs substantively from what a smoker inhales. In response, Jennifer Jinot and Steven Bayard of the EPA argued that the agency's conclusions were based on "a comprehensive analysis of the total weight of evidence, in accordance with the Agency's risk assessment guidelines." They stated that the EPA assessment had been subjected to public review twice, and to review by an independent scientific panel of 18 experts.

The primary argument was whether or not environmental tobacco smoke is related to lung cancer. Also at issue were respiratory effects. As arguments in favor of asserting that environmental smoke is hazardous, Jinot and Bayard cited these findings:

- Environmental smoke is chemically similar to mainstream smoke; although they differ somewhat, they contain the same toxic agents and are considered similar for the purpose of qualitative identification of hazards.
- The EPA's statements about lung cancer were based on the well-established association between active smoking and lung cancer, and the fact that biological measurements show that nonsmokers take in and metabolize tobacco smoke components. Additionally, studies with animals indicate that tobacco smoke is carcinogenic. Also, 30 epidemiological studies of environmental exposure and lung cancer among nonsmoking women in eight countries indicated that exposure is linked to lung cancer.
- Evidence clearly indicates that exposure to environmental tobacco smoke causes respiratory problems in children.

Jinot and Bayard commented: "The EPA analysis concludes that the overall consistency of positive responses in numerous studies of different design from many countries, consistently positive and statistically significant exposure-response relationships, and consistently higher risks in the highest exposure groups provide sufficient evidence that the risks of both lung cancer and noncancer respiratory effects from ETS exposure are real and cannot be ignored."

The debate continues with vehemence in many arenas. In September 1997, columnist Robert J. Samuelson complained in *The Washington Post* that in all of the sensitivity directed toward minorities recently, one minority had been overlooked: smokers. Samuelson argued, "The debate over cigarettes has been framed as if smokers are the unwitting victims of the tobacco industry." According to Samuelson, smokers are treated as if they "lack free will and, therefore, their apparent desires, opinions and interests don't count." His article also called the health risks from environmental tobacco smoke into question.

Samuelson's column drew sharp response from former U.S. Surgeon General C. Everett Koop, viewed by many as the guardian of the nation's health for nearly two decades. "As I understand it," Koop began, Samuelson was claiming that respected newspapers and educational institutions "have been duped by a group of anti-smoking zealots and public health loonies." He criticized Samuelson's lack of data to support his counterclaims about environmental tobacco smoke. "I agree that smokers have rights," Koop stated, "but the right to harm others is not one of them."

Going Smoke-Free

Just what happens when smokers are forced to go smoke-free at their workplace can be a surprise to all involved. In 1989, the author of this book and Maxine Stitzer at the Johns Hopkins University School of Medicine ran a study investigating that question as the Hopkins medical institutions converted to smoke-free workplaces. As the ban was being implemented at what was then the Francis Scott Key Medical Center, we followed 34 smokers who were no longer allowed to smoke at their work stations. Before the ban, some of them found places in their work areas where they thought they could get away with smoking undetected, such as closets or unused rooms. One smoker, fearful of any restrictions, threatened to set up her office outside her window on an adjoining rooftop, where she could smoke. (It was easily accessible through a window, she explained.)

The implementation of the smoking ban brought a sudden decrease in the amount of tobacco they consumed. However, few of those in the study followed through with their threats to smoke in secret places inside the buildings. One worker explained that once the time came, it simply seemed like too much bother. The smoker who had considered setting up an open-air rooftop office perch found that she was able to cope, despite some discomfort. Tobacco exposure, as measured through breath carbon monoxide and through nicotine levels in smokers' saliva, declined. Overall tobacco exposure, as measured by the nicotine metabolite cotinine, decreased by 15 percent, but the decline was not statistically significant. Smokers used an average of four fewer cigarettes per day. The smokers we assessed experienced abstinence effects consistent with their reduced exposure to tobacco, including cravings for cigarettes, urges to smoke, difficulty concentrating, increased eating, and depression symptoms.

To determine whether the smokers engaged in compensatory smoking, we not only measured biological indicators of exposure to nicotine, but also counted and weighed their cigarette butts, which they dutifully collected in little plastic bags every day. They kept butts smoked at different times of the day separated in different plastic bags, so that we could see whether they smoked more in the mornings or before work (some worked on night shifts). This allowed us to determine whether their smoking patterns shifted, or whether they otherwise compensated for not being able to smoke as they worked.

We desiccated the butts to a consistent 20 percent humidity level, to compensate for the Baltimore, Maryland, summertime humidity that frequently reached 80 and 90 percent. We then counted, weighed, and measured the length of the butts; cigarette butt weight was an indirect measure of the intensity of smoking. Butt length, like butt weight, also indicated the amount of material left unsmoked in the cigarette. Butt

Butt—Butt—Butt—

204,544	number of cigarette butts picked up in 1995 California statewide beach cleanup
17%	California population that smokes
about half	proportion of California beach debris consisting of cigarette butts
foam plastic	second most common California beach debris item
junk food wrappers	third most common California beach debris item
about 8,000	yearly reports of toxic exposure to tobacco products among children younger than age 6 in the U.S.
146	reports of nicotine poisoning by children under age 6 in Rhode Island in an 18-month period
high	likelihood that cases of cigarette-butt ingestion by children are underreported
high	likelihood that a nicotine-poisoned child lived in a home where adults used tobacco in the presence of children

length also provided an indirect measure of toxicity, since tar and carbon monoxide delivery increase logarithmically as a cigarette is smoked to a shorter butt length.

We found that the smokers did not seem to compensate to an appreciable degree for the smoking restrictions. They did not smoke the butts down farther and did not load up on nicotine to any noticeable extent

Bans Roll On

up to 25%   extent to which a workplace smoking ban can reduce smoking

9%   percent of workers, unable to smoke at work, expressing a strong need to smoke at work

71%   percent of workers, unable to smoke at work, expressing a mild or occasional need to smoke at work

---

before coming to work. Instead, they reduced their overall exposure and, for those hours at work, put up with the discomfort of abstinence symptoms. We had anticipated that many of the participants would want to use the workplace restrictions as a reason for quitting smoking, particularly since the hospital offered smoking cessation help for any employees who wanted to quit. To our surprise, not one of our study participants quit smoking as the ban was implemented, or for six months following.

Another surprise came when we sought a comparison group of smokers who were still allowed to smoke at their hospital worksites. As we called hospitals within a radius of about a hundred miles, we found that all either had implemented or would soon implement smoking restrictions. When we had started our study at Francis Scott Key, few hospitals in the area had smoking restrictions; within just a few months, virtually all of them either were restricted or were soon to be restricted. The policy changes had come so quickly that to populate our comparison group, we had to look at work settings other than hospitals to find smokers who weren't undergoing smoking restrictions.

Compromises

Some tobacco-control experts have a dream that someday the world will be free of tobacco. They know it is not a realistic dream. Some smokers also have a dream that someday the nonsmoking world will quit hassling them about their smoking. Imposing draconian restrictions on smokers has been known to trigger backlash among some opponents. For instance, the 1998 implementation of a smoking ban in bars in California was "the last straw" for many smokers. Smoking already had been banned in California restaurants for some time; bars were among the last places that smokers could smoke in public settings.

California's bar ban went into effect at midnight, January 1, 1998. Many smokers celebrating in California bars that night did not compliantly extinguish their cigarettes and go on about their merrymaking. Instead, some bars set out cash jars so that customers could contribute to paying the fees for those who would be fined for violating the prohibition. Many smokers just kept smoking, with the full blessing of the proprietors of the establishment. Enforcement was sporadic, at best. Some bar owners set up separate smoking areas where customers could light up but employees could not go. Within weeks, the lower house of the California legislature, the Assembly, considered rescinding the ban, which even some nonsmokers saw as being too restrictive. A Californian supporting the ban remarked on a radio program that the Assembly was attempting to respond to its constituency on the issue with amazing speed; if only it would respond that quickly on issues such as health care, he noted dryly.

Californians' responses to the state's groundbreaking smoking restrictions reflected either love or hate, but rarely anything in between. *San Francisco Examiner* columnist Rob Morse wrote a column entitled "Emission Control Problems," in which he described imaginary restrictions on smoking in cars (which, by the way, were never proposed). Those caught smoking in their cars would have their drivers' licenses revoked and returned the next year "with an even worse picture." As if that weren't dire enough, he offered the prediction that automobile ashtrays would be required to be filled with change, not with cigarette ashes, and would be known as "cashtrays." He concluded, "Now about those smoking and no-smoking lanes . . ."

The battle lines over smoking restrictions have been drawn in many places and many venues, such as these:

- In Sierra Vista, Arizona, the city council considered a ban on virtually all indoor smoking. In response, smokers started an initiative requiring restaurants to have nonsmoking sections, but otherwise nullifying the rest of the proposed ordinance. The National Smokers Alliance of Alexandria, Virginia, helped smoking advocates collect signatures to put their proposal on an upcoming ballot. Health advocates elsewhere in Arizona fought six attempts by the state legislature to ban local smoking ordinances.
- About half of county prisons nationwide have no-smoking policies either in place or in planning. Eleven state prison systems and some federal prisons also ban smoking among inmates. Prison authorities' biggest fear from implementing the bans, they say, is the possibility of riots.
- In October 1997, the tobacco industry agreed to pay some $300 million to establish a medical foundation to study illness related to tobacco use. The agreement settled a lawsuit in which about 60,000 nonsmoking airline flight attendants sued the tobacco industry for $5 billion, claiming illness from breathing secondhand smoke on airplanes. The settlement came a month into the defense's presentations before a jury.

There may be no simple resolution to the conflict between smokers and those exposed to their smoke. Pro-tobacco rhetoric notwithstanding, the health hazards of environmental exposure to tobacco smoke are accepted widely throughout the scientific and medical communities. And even if tobacco smoke weren't toxic, many nonsmokers (and some smokers) find it so noxious that they prefer that it be restricted, toxic or not.

On Risk

Risk analysis specialist W. Kip Viscusi noted in 1992 that "Risk taking is an inescapable feature of our lives." He characterized smoking as "an individual risk-taking activity," independent of involuntary exposure to secondhand smoke. "If these choices are informed and have no adverse effects on society, there would be no efficiency rationale for regulating this behavior."

However, the choices smokers make do affect others. Not all smokers use tobacco in a way that preserves others' right to clean air. As Viscusi explained, "Smokers may take some . . . risks into account, particularly for family members, but are not likely to undertake actions that are fully optimal from a society perspective."

Hence comes the need for government regulations protecting those who otherwise are placed at risk by others' use of tobacco. In civilized society, as adults we do not inflict our detritus on others. Our cars have mufflers and smog-control devices. Our waste-water goes into a sewage system. We cough behind our hands. We do not double-dip our potato chips. We do not spit in others' food, or anywhere else that others might be, for that matter.

And we should not expect others in our immediate vicinity to breathe our smoke.

CHAPTER 8

E verything costs.
"Some people just let a cigarette sit" after they light it, she says, indicating disapproval. "Some people who smoke two or three packs a day, they'll light a cigarette and they'll let it burn." Ever since she started allotting herself only ten cigarettes a day, she's been more aware of the "value" of those ten cigarettes in her smoking life.

"I don't just light one and let it burn. If I'm gonna light it, I'm gonna smoke it, because I don't want to waste it, because now I'm smoking less. That would take away one of the cigarettes that I'm allowed to smoke during the day. I certainly can't afford to waste a cigarette.

"If I smoke half a cigarette and then I put it out, then, guess what, I've only used of half of something that I could've had a whole of if I'd waited until I had time to really smoke it. And now I have one less cigarette that I'm allowed to smoke during the day."

Because everything costs—and in her internal economic budget through which she allows herself only so-many cigarettes per day, wasting a cigarette she could've smoked is like wasting money, only worse. It's wasting what she could've experienced and felt.

═══════

Trade-Offs

E verything costs, and everyone pays. Often, we don't realize we are paying, and often we don't realize the thresholds of what we will pay. But we do pay.

Let's imagine that someone gives us a choice: Either we can have $100 now or we can have $100 one week from now. Most of us would choose to have the money now. But if we are offered $100 in six years, or $200 in eight years, most of us would take the $200. Our preference has been reversed as a function of the long time it would take to get either amount. Valuation changes as a function of time, and this then affects our preference.

"People figure it's such a long ways off, they may as well get the larger amount," explained researcher Warren Bickel, who has helped to do pioneer work in what is called the *behavioral economics* of tobacco use. He and other scientists have applied and expanded economics principles to the field of drug abuse.

The currency or commodity can also be tobacco, rather than money. For example, a smoker can have a cigarette now or a better chance for

The Cost of Being a Smoker

$3,650	cost of purchasing a pack of cigarettes a day for 5 years, not including inflation or lost accumulative interest
$310 – $520	annual cost of extra cleaning bills for a smoker vs. a nonsmoker
$3,963	added amount a middle-aged smoking male pays for a standard term life insurance policy over 5 years, compared with a nonsmoker
42%	men below the U.S. national poverty level who smoke
30%	women below the U.S. national poverty level who smoke
4% – 8%	decrease in income earned by smokers, compared with nonsmokers
1 in 3	chance of dying of smoking-related disease

good health in 35 years. While good health today may seem worth more today than a cigarette, good health in 35 years seems less valuable. Also, a smoker can pay about $2 for a pack of cigarettes now, or instead can someday use the money he or she would have spent on cigarettes to buy a luxury car or even buy a house. Given these choices, many smokers still opt for the cigarette now. This is one of many situations in which we "sell" a valuable asset today for very little, if the "delivery" seems far in the future.

These types of now-or-later choices are evident in statistics on tobacco use during pregnancy. An Australian study found recently that

21 percent of pregnant smokers ignored the risks of smoking and continued using tobacco during pregnancy. Some 45 percent of pregnant Australian young women who smoked continued to use tobacco while they were pregnant. The 1997 study by Patrick P.L. Wong and Adrian Bauman examined more than 85,500 births in New South Wales in 1994 and found that the women who continued to smoke risked obstetric complications, premature birth, and low-birth-weight babies. Their offspring also were more prone to respiratory infections, asthma, growth retardation, cognitive deficiencies, and even death.

Many of these women, no doubt, planned to quit smoking if they became pregnant, and many probably intended to quit all the way through the pregnancy. As one 38-year-old mother of four told the researchers, "I worry about it, but obviously I can't worry too much. I've always been aware of the risks, but I just haven't really taken it too seriously."

Such logical inconsistency makes sense—sort of—when viewed through the lens of behavioral economics. "The farther you are away from events, your behavior will seem more self-controlled and rational," Bickel noted. The farther away we believe that a cost will come due, the smaller the cost seems. As our time horizon changes, the costs and benefits of our present values change.

For example, if we have a looming deadline, such as a report due first thing Tuesday morning, we might decide on Sunday night that we need to get up very early Monday morning to get a good start on the work. The cost of getting up early seems less on Sunday night than it seems early the next morning, when we feel too tired to get out of bed. "We'll set the alarm to get up early and get a lot of work done," Bickel explained, "but in the morning we hit the snooze alarm." The present value of sleep has increased as the cost of writing the report has increased. "It's not a lot different than a smoker deciding to stop, but then walking down the street and seeing someone smoking a cigarette, and not stopping smoking."

Unlocking the secrets of human behavioral economics might help lead to better smoking cessation treatment. Adapting and expanding on principles from the larger and older field of microeconomics in the 1970s and 1980s, behavioral economists have identified some of the reasons that smokers make the choices they make.

If there are two cigarettes of equal amounts of nicotine but different cost, the buyer's income could affect preference. If a less desired cigarette costs one-half or one-third of the more desired cigarette, persons of lower income tend to choose the less desired cigarette and persons of higher income to choose the more desired cigarette. Having a choice between these two cigarettes might result in this scenario: If our income increases, we would use the more desirable cigarette that costs more and would use the less desirable cigarette increasingly less. As we get wealthier, we would be buying more expensive cigarettes, or perhaps cigars, and smoking fewer generic cigarettes.

How does this analysis relate to stop-smoking treatment? Bickel explained that in economic terms, tobacco products might be considered preferable to nicotine replacement products that do not provide quite the same experience as smoking. The irony should be apparent, since nicotine replacement products undergo exhaustive development and testing before being marketed, and are "inferior" only in that most smokers perceive them as less desirable than tobacco. Work by Bickel and his colleagues indicates that smokers do tend to perceive nicotine replacement products as what economists call an "inferior good," or one for which "demand" falls as "income" rises.

"What this suggests, if this is representative of how nicotine replacement operates in the real world, is that the only way to make it a viable alternative to cigarette smoking is to radically increase the price of cigarettes and to radically decrease the cost and increase the accessibility of nicotine replacement," Bickel explained. "You can demonstrate that people consume fewer cigarettes when they have nicotine replacement, but the important question is, under what conditions will they use the products?" Smokers are not electing to quit en masse, and only a minority employ the harm-reduction strategy of using nicotine replacement. Perhaps smokers would attempt to quit in greater numbers if nicotine replacement were available in smaller packages that were available at convenience stores, or if the price differential favored the replacement product over cigarettes. "Those are the conditions under which you would engender the nicotine replacement option," Bickel noted.

In laboratory research with human smokers, Bickel found that some smokers would perform considerable "work" (i.e., tasks in the study) for a puff on a cigarette. His all-time hardest-working research subject was willing to engage in 2.5 hours of computer tasks for two puffs on a cigarette. Amazingly, the same subject could have waited only 3 hours for that segment of the study period to end, without engaging in the computer work, and then could have gone outside and smoked.

"This suggested to me that indeed the temporal horizon into which cigarettes are integrated over time is pretty small," Bickel said. That temporal horizon apparently is one of the key factors at work in the decision to use tobacco.

To understand how this process works, it helps to consider some monetary examples. If, for instance, we are given the option of taking $1,000 now or $1,000 a week from now, the choice is fairly predictable. But what if we can have $980 now or $1,000 in a week? Researchers find that as they manipulate the amount of discount that someone will accept, they can detect a direct measure of how much a person will discount $1,000 in a week. The same question can be posed and the same calculation made at many different time points to provide a picture of the rate of discounting. "It's almost a psychophysical procedure," Bickel explained, comparing it to measuring such physiological phenomena as heart rate or brainwaves.

"When people keep doing things that are bad for them, it may be because their temporal horizon is short," Bickel stated. "Things beyond a certain point don't matter to them." If comparable questions are asked of heroin abusers and of research subjects who do not abuse drugs, nondrug users will perceive that the $1,000 will lose 60 percent of its value in about five years; for heroin addicts, $1,000 loses the same amount of perceived value in one week. When the commodity becomes heroin, no amount of heroin tomorrow is worth any amount of heroin

Up in Smoke

$2	amount of hidden health costs associated with each pack of cigarettes sold
$47 billion	annual loss from sick leave and loss of workers to death related to tobacco use in the United States
33 cents	amount per pack of smokers' net loss to society
25 cents	amount per pack of net loss to society related to secondhand smoke
24 billion	number of packs of cigarettes purchased in the United States annually
$50 billion	annual U.S. cost of health care for smoking-related illness
$47 billion	annual U.S. cost of lost productivity and forfeited earnings due to smoking-related disability
one-third	proportion of world's cigarettes used for smuggling
quadrupled	increase in tobacco smuggling from 1989 to 1993
$16.2 billion	annual amount governments lose worldwide because of tobacco smuggling
154,000	number of fires started each year by tobacco-related smoking materials
$416 million	annual damage from fires caused by smoking-related materials

today to a person dependent on heroin. Behavioral scientists studying smoking now are attempting to examine similar questions in relation to tobacco and alcohol use.

The work of Bickel and others has shown that virtually all self-administration of drugs, whether by humans smoking tobacco or by laboratory primates self-administering cocaine, fits the same shape curve. When the amount used is plotted against the timing of administration, the resulting graphics look similar, whatever the drug. The shape stays constant, although its parameters vary according to the specific drug. This fact has given behavioral economists an empirical, measurable model on which to base assumptions, "which is a useful contribution of psychological science to economics," Bickel added.

Psychologists and others engaging in this work can learn much from economics, particularly its concepts and terms. "Other concepts that may be particularly useful include the notion of reinforced interactions," Bickel explained. "In economics, reinforcers can interact as *substitutes*. For example, if we increase the price of seeing a movie in a theater, its consumption will decrease, but there will be an increase in the use of videos even though their price remains constant." Another type of interaction is the *complement*, in which consumption of soup would decrease as the price of soup increased. The consumption of soup crackers would also decrease, even though their price remained constant. Lack of interaction would indicate independence; for example, the price of a theater ticket would not affect the consumption of soup crackers.

"This way of thinking provides a useful framework for understanding events that interact with cigarette smoking," according to Bickel. An increase in cigarette prices will decrease coffee consumption along with tobacco consumption, for example, but an increase in coffee price will not affect cigarette consumption. "Those events can be confusing when the effects occur together," he admitted. Studied carefully, however, this behavioral economics approach can "provide a conceptual system for studying how tobacco interacts with other drugs, and also with social interactions." Additionally, it raises valuable questions and provides a useful framework for quantification.

"If you could create an analog of an ideal smoking treatment situation, you would like to have a world in which you would have few complements for smoking, and many pro-social substitutes," Bickel stated. "You can test that in a laboratory and test it in the real world." Behavioral economics provides a conceptual system for understanding how elements go together. An example of the kind of statistical puzzle that behavioral economics can help solve is the use of multiple drugs sequentially, concurrently, or both. Some drugs are used together, some are used in lieu of each other, and some are used together even though there is no relationship between then. "How do you put that quagmire in functional categories?" Bickel asked. "This is a potential way of doing that."

Many baffling aspects of human behavior, including smoking-related behaviors, can be studied through the eyes of behavioral economics. One of the most puzzling, the "loss of control phenomenon," is readily analyzed through this means. It is a phenomenon many of us know well: We make a decision to do something, but when we are confronted with the reality of it, we find ourselves choosing the direction we had not planned to take. Bickel and others have studied this "preference reversal" situation in terms of "discounting." In other words, what looks like a good idea from a distance of time no longer seems as appealing when we approach the situation. The closer we are to paying a price, the higher the present value of the price. Our resolve to diet was strong a week ago, but instead of choosing to lose weight today, we find ourselves eating an ice cream cone. Such a choice can feel as if it is out of our control. When the concept is applied to drug use, it becomes a diagnostic criterion for the misuse of many substances, including nicotine. Researchers have not yet determined whether behavioral economics can be applied directly to smoking cessation treatment, to lengthen a smoker's horizon or to manipulate preference reversals.

"Studies have looked at time perception of drug-dependent people," Bickel explained. "People in treatment have a longer time-perspective than those not in treatment. Hopefully that means that treatment can induce a longer time perspective." It is also possible that persons with a longer time horizon are more able to value treatment. For many substance abusers, including many smokers, any inducement toward a longer time horizon would need to be attractive indeed. Bickel found that when most people think about the future, they view ahead an average of nine years. Heroin addicts, on the other hand, view the future as being four weeks away. "And those are the ones who are in treatment," Bickel explained. "God knows what it's like out in the street."

Adolescent alcohol and drug use presents a particularly poignant example of this concept. "Do adolescents get bumped into drugs when they should be expanding their temporal horizon?" Bickel has wondered. If so, maybe adolescent drug use is a different sort of problem than it previously has been considered to be.

One of the leaders in applying behavioral economics to substance use is Steven R. Hursh of the Johns Hopkins University School of Medicine and the Science Applications International Corporation. In the early 1970s, as he started his career in science at Walter Reed Army Institute of Research, he noted inconsistencies in baboon behavior. Scientists taught baboons to give themselves heroin and studied the primates' patterns of use as a way to develop a possible model for human use of heroin. Hursh noted that if he wanted the baboons to respond at a higher rate, he had to make them "work" harder to get the drug. "This seemed counterintuitive," he recalled. "It turns out, though, that it wasn't at odds with what economics would predict."

To understand the baboons' heroin use, he referred to the concepts of *elastic* and *inelastic* supply and demand of commodities. The demand for a product that we need to buy, whatever the cost, is inelastic. "This concept opened the door to realizing that the factors that control our overall output of performance for a commodity can be better understood from an economic perspective than from a simple reinforcement perspective," he explained. "Reinforcement is part of a process in which the subject is trying to defend some level of consumption, and that consumption has some benefit, such as enough food to maintain health, or enough drug to ward off withdrawal."

Some commodities do not maintain that kind of inelastic demand. For instance, monkeys "working" to obtain nonnutritive saccharine will work for it when the flavoring is easy to get, but when the price is raised to the level of a food price, the monkeys do not increase their work output. This is also true of laboratory rats taught to press a bar to obtain food, or laboratory pigeons taught to peck a certain spot to obtain grain for food. If the task is made more difficult, the animals will reduce their rate of response.

"This led to a study that showed that this occurs because rats and pigeons are kept in an open economy, in which they don't get most of their food from the experiment," Hursh said. "So the food from the experiment is a luxury good. But if you study monkeys or rats in an environment in which the only food they get is from the experiment, they show that the demand is inelastic."

"All of this is part of a unified process, but you have to understand the nature of the economy that they're working in. If the food outside substitutes for food within the experiment, the demand for experiment food is elastic," Hursh explained. The continuum between elastic and inelastic demand for tobacco and other consumed substances can be controlled and manipulated. By altering "price" (cost and availability), one can make a substance so expensive in money and in effort that regular users will use it less and less frequently.

This approach has been applied, with mixed success, to the problem of adolescent tobacco use. Tobacco use by minors is illegal virtually everywhere in the United States, yet sizable percentages of youth use tobacco. Governments and municipalities have tacked substantial taxes onto tobacco to make it less accessible to young persons. The theories and evidence indicate that this approach ought to work in reducing tobacco use by children and adolescents. Even so, the results of altering prices or decreasing accessibility of tobacco have been mediocre.

A 1997 report in *The New England Journal of Medicine* by a team led by Nancy Rigotti of Massachusetts General Hospital indicated that tightening enforcement of laws prohibiting tobacco sales to minors did not reduce consumption by minors. Rigotti and colleagues examined enforcement of tobacco laws in three communities by having minors who worked for the investigators attempt to purchase tobacco from retail vendors. In addition, anonymous surveys of more than 22,000 students in grades nine through twelve assessed tobacco access and use among adolescents.

At the start of the study, some 68 percent of 487 vendors sold tobacco to minors. When local health departments distributed written information about tobacco sales laws and began testing and enforcing compliance (these tests were separate from those conducted by the study investigators), compliance increased to 82 percent, which did not quite meet the goal of more than 90 percent. Concurrently, the same measurements were taken in communities where law enforcement agencies did not go to any extra lengths to enforce tobacco sales regulations. Those communities maintained a 45 percent compliance rate. The change in compliance rate was significantly different in those communities with more rigorous enforcement, but the change failed to affect minors' access to tobacco. Teenagers reported only a small decrease in their perceived ability to purchase tobacco in the communities where enforcement was enhanced, and they reported no decline in tobacco use.

The problems contributing to the disconnection between tobacco-control efforts and reported use are probably manifold. The goal of reducing young persons' use of tobacco is a hallmark of tobacco-control legislation and negotiation, yet evidence does not support the efficacy of the standard approaches. As Rigotti and colleagues stated, "Empirical demonstration of benefit is critical to justify the resources being expended on this new effort." In addressing the question, they concluded that typical tests of compliance with tobacco laws tend to underestimate young persons' access to tobacco. Potential reasons include the possibility that underage smokers attempting to purchase tobacco may lie about their age or use false identification, or may enlist an older person to buy tobacco for them. Also, store owners might sell tobacco only to youths whom they already know. Teenagers also might buy tobacco in neighboring communities that enforce tobacco sales regulations with less rigor.

Defining Values

Howard Rachlin, a State University of New York at Stony Brook researcher of human behavior, wrote that behavioral economics sometimes is seen as the "study of error—of deviation from rational behavior." Such " 'psychological' good" as approval, satisfaction, and avoidance of negative feelings can translate into "subjective value," and value is precisely what human decision making seeks to maximize. Which is to say, if someone places greater value on the sensations associated with smoking than on physical health, smoking may have more value for the person than jogging, or than quitting smoking.

Rachlin stated that he prefers to look at behavior (such as smoking) from within the framework of the overall pattern in which it is performed. Someone looking at smoking from Rachlin's point of view might see smoking not as an isolated act, but as one of several mechanisms for maintaining emotional equilibrium. He equated what we might call rational behavior with self-control, and equated irrationality with impulsiveness. The same act can be either rational or irrational, depending on its context. For example, smoking a cigarette is neither rational nor irrational in and of itself. But the smoking of a cigarette by someone who has quit smoking would be considered irrational, while the smoking of a cigarette by someone who is cutting down on smoking as a prelude to quitting would be considered rational. He insisted that the study of what might be considered anomalous behavior should not end with its mere discovery and classification, but should be examined in terms of its "functional base in the environment."

An example of this "functional base" could be a behavior that does not serve its obvious purpose. A "wider interest" or long-range interest, such as maintaining relationships by smoking with friends, can interact with a more immediate interest, such as the pleasure of smoking and the alleviation of withdrawal symptoms, to dominate an interest that is somewhat mid-range, such as quitting smoking to improve health. Personal maximization based on one alternative or another "is strictly relative and conditional," Rachlin stressed. "We cannot presume to know what an organism (including ourselves) values. Rather, value must be defined by choices."

1,000 Tons, and What Do You Get?

$2.6 million	added economic value from 1,000-ton increase in tobacco consumption
$5.6 million	estimated direct costs of medical treatment related to consumption of 1,000 tons of tobacco
$29.8 million	total cost of death and illness related to consumption of 1,000 tons of tobacco
minus $27.2 million	benefit-cost difference of consumption of 1,000 tons of tobacco
650 – 1,300	average number of deaths per 1,000 tons of tobacco consumed
25 – 30 years	time lag between processing of tobacco and the related deaths

(Barnum, 1994.)

The choice that most of us make to have something now rather than something else of value some time in the future may reflect our impulsiveness, but it is not an anomaly, Rachlin insisted. He refused to treat such choices as deviations from the standard rules; instead, they are behaviors to be studied, so that more accurate rules can be discovered. Turning this approach inward toward the individual ex-smoker's attitude might involve not labeling a lapse back to smoking as an anomaly, but rather as a situation to examine in view of the individual's values and choices. Why does someone who knows the value of a healthy diet consistently overeat? Why does someone who knows the dangers of smoking continue to smoke? Those questions are not examinations of anomalies in the stream of rational behavior, but are case studies that provide an opportunity for understanding the behaviors within their larger framework.

"Self-control is difficult," Rachlin stated, citing two reasons. First, as behavior patterns become more abstract and extended over time, perceiving them is more difficult. The common outcome of patterns is weaker and more difficult to spot. Second, even if we can perceive them, their elaborate patterns require maintaining a wider view of the events, or a broader "temporal horizon" in the presence of immediate "reinforcers" such as the sensation of smoking, or the relaxing or stimulating effects of nicotine. A successful quitter who indulges in one cigarette faces what Rachlin called "an awful conflict," with haunting questions: Is the immediate reward worth risking slipping back into regular use of tobacco? The smoker thus, in a sense, is "tossing the dice with regard to the future," to apply Rachlin's terminology.

What will happen? Will the lapse cigarette be the last, or will the ex-smoker become a smoker again? What is behind the door in the arena, the lady or the tiger? We won't know, Rachlin explained, "until future behavior settles in."

The economics that apply to an entire population of smokers might not explain the behavior of an individual smoker. Economics may need to be applied differently at different levels of analysis. It might not be accurate to assume that the methods by which a behavioral economics researcher might analyze the behavior of a group of individuals would be useful in studying the economic impact of a cigarette tax or cigarette regulations on a larger group chosen randomly from the population at large. And what is valid for one segment of that population, such as adolescents, might not be the case for the rest of the population.

Analysis of cigarette taxes and cigarette consumption in all 50 states in the United States over a period of 40 years indicated that cigarette taxes effectively reduced cigarette use, but that federal taxes had more impact than state taxes, presumably because federal taxes provided no advantage for "bootlegging" cigarettes across state lines. The report also noted that cigarette consumption declined as health warning labels were added to cigarettes. Reducing consumption by large amounts required large increases in taxes, Kenneth J. Meier and Michael J. Licari of the University of Wisconsin-Milwaukee concluded. Increases in excise taxes are associated with reduced tobacco consumption, even when other factors are considered statistically. Interestingly, they discovered that the rise in tobacco-related health concerns of the 1960s actually diminished the effect of taxes, because "individuals have reasons other than economic ones to stop smoking." They recommended substantial increases in federal excise taxes to reduce the demand for cigarettes.

Harm Reduction

To quit or not to quit? That is a question smokers face. To help mitigate the risk of tobacco use, some experts have developed an approach called *harm reduction*. They view it as an alternative that might help save lives, even if purists might not approve. Millions of smokers who find it too difficult to quit smoking completely might consider other alternatives. Experts have debated the benefits and risks of helping smokers cut down how much they smoke, or find safer ways to smoke, as a way to reduce harm among those unwilling or unable to quit smoking completely. Some 16 million smokers try to quit every year; of these, only about 1.2 million succeed.

Strategies for harm reduction can include the extended use of nicotine replacement such as nicotine gum or nicotine patches, switching to alternative nicotine delivery mechanisms such as smokeless tobacco, or even using cigarettes that are less toxic, such as a smokeless cigarette or other smoking device designed to minimize harm. Another strategy for harm reduction is smoking fewer cigarettes per day.

Even though smoking prevalence has leveled off in the United States, tobacco experts are faced with the problem of finding more options for hard-core smokers who are not likely to quit. The approach could "complement, but not replace" efforts to prevent smoking onset and aid current smokers in their attempts to achieve complete tobacco abstinence, said Johns Hopkins University researcher Jack Henningfield, who has publicly debated the issue with other experts.

"Medical and public health experts are considering reducing smoking as a goal of intervention for those smokers who are presently unable or unwilling to completely quit smoking," Henningfield said at a 1997 symposium where harm reduction was debated. This "exposure reduction" strategy could lead to eventual smoking cessation, but the goals are more immediate. Smokers who are not yet ready to quit may gain some health benefits from reduced exposure as they cut down the number of cigarettes. This would reduce their exposure to the harmful carcinogens and tars in tobacco smoke. In addition, reduced exposure to smoke can raise their chances of quitting smoking.

Results from three teams of researchers who presented findings at the symposium indicated that exposure reduction might help reduce the death and disease associated with tobacco use:

- Smokers who reduced their smoking were more likely to quit than were smokers who did not, according to a research team at Oregon Health Science University at Portland. They found that long-term users of nicotine gum were able to cut down their smoking and, eventually, their gum use.
- Researchers conducting community-based smoking cessation efforts in Vermont found that smokers were able to reduce their smoking and to maintain the lower levels for a long period of time. Those who cut their smoking by at least half had a slightly greater tendency to attempt to quit smoking, but the effects were not statistically significant.
- Smoking fewer cigarettes per day offered health benefits for smokers, according to researchers from Göteborg, Sweden, who examined the influence of smoking reduction on cardiovascular risk factors. Their study investigated the impact of smoking reduction on those smokers who know that they should quit for health reasons, but who have difficulty quitting. Men and women who reduced their smoking by half experienced a beneficial effect on cardiovascular risk factors.

One researcher, David M. Burns of the University of California at San Diego School of Medicine, estimated that if 100,000 two-pack-a-day smokers reduced their smoking to fewer than ten cigarettes per day at age 50, they would cumulatively gain an additional 284,000 years of life. This is based on an estimation model for calculating public health benefits from smoke exposure reduction.

Another approach to reducing tobacco use is to limit the amount of nicotine, which is believed to be tobacco's primary psychologically effective (*psychoactive*) agent as well as its addictive component. Proponents of this approach recommend gradually reducing the legally allowed amount of nicotine in cigarettes until the amount reaches a virtual zero. The rationale is based in findings that nicotine is the primary tobacco constituent that drives the intensity of smoking. In an effort to get all the nicotine possible, smokers will commonly puff more, inhale more, and hold the smoke longer in their lungs when they smoke cigarettes with lower nicotine content. Recent research has shown that this particular smoking style carries the additional health risk of a type of cancer seen in those who inhale very deeply, as is common in the use of low-nicotine cigarettes. Also, those who smoke low-yield cigarettes tend to smoke more cigarettes and to get more puffs from each cigarette. All of these actions combine to compensate for the lower nicotine available from each cigarette.

Some experts claim that it is unlikely that gradually lowering the amount of nicotine in cigarettes will reduce overall smoking. It may, in fact, increase smoking as nicotine-dependent tobacco users strive to maintain the blood levels of nicotine required to avoid withdrawal.

The Bottom Line

U.S. News Online, an Internet production of the weekly magazine *U.S. News & World Report*, related this tale by Matthew Miller: A tobacco-control crusader asserted at a Canadian conference in the 1980s that the United States was spending more than $50 billion per year in smoking-related health costs and lost productivity. After the presentation, an economist insisted that the amount couldn't be right because it didn't account for savings in Social Security and private pensions for smokers who had died prematurely. The crusader snapped back, "You know that, I know that, and the tobacco industry knows that. The tobacco industry can't *say* it, because it means admitting they're killing their customers. Are *you* going to say it?"

It is, indeed, being said these days. A frequent consideration in the overall picture of the economics of tobacco use is the down-the-road cost of using tobacco. For the individual smoker, health risks decrease when tobacco use stops, with the decrease growing in magnitude the longer the ex-smoker stays abstinent. Not only has the ex-smoker saved money, but the smoker's lower tobacco-related health costs are savings as well. It sounds so simple.

But it isn't. A recent report by Jan Barendregt and colleagues of Erasmus University in the Netherlands studied the effects of smoking on health care costs. They compared a mixed population of smokers and nonsmokers, a population of smokers, and a population of nonsmokers (*population* being a research term indicating the larger group under study, of which the research subjects are a representative sample). They used a "dynamic" statistical method to estimate the effects that smoking cessation would have on long-term health care costs. They determined that at a given age, health care can cost as much as 40 percent more for smokers than for nonsmokers.

The picture, however, was not that simple. Increases in longevity contributed substantially to the overall increase in long-term health costs in a society of nonsmokers. Smokers have a life expectancy at birth of 70 years for men and 76 years for women; nonsmokers' life expectancies are 77 years for men and 82 years for women. More nonsmokers than smokers live to old age. For example, 50 percent of nonsmoking men are still alive at age 80; 21 percent of smoking men are still alive at the same age. Younger smokers have higher health care costs than their nonsmoking counterparts, but over their respective life spans, nonsmokers' health costs are greater because they live longer. The health problems nonsmokers face in old age may be unrelated to their status as a nonsmoker.

Ironically, a society could greatly diminish its health care costs by not encouraging public health and by allowing disease and death to run rampant in people of all ages. Health care costs would be low, but quality of life also would be low. As with the choices made by the individual smoker, a society's choices reflect its values. Barendregt and colleagues posed this question: "With respect to public health policy, how important are the costs of smoking?" Their response was that public health policy should be concerned with health, and that "whether or not smokers impose a net financial burden ought to be of very limited importance." Whatever the bottom line, they said, "Smoking is a major health hazard." Consequently, smoking should be discouraged. Period.

The Smell of Success

Perhaps the least predictable extension of private values into the public breathing space has been the recent explosion in cigar popularity. Did the trend start with the photo of U.S. President Bill Clinton chomping on (but not inhaling) a cigar? Was it the many heroes and antiheroes who puffed on cigars in the movies? Was it the parade of famous faces on the cover of cigar magazines? Was it the magazines themselves?

The long-term cigar slump that half-heartedly peaked in the 1970s reversed dramatically in the mid- to late 1990s as cigar use became a trend, even among the health-conscious. The cost of cigars increased along with the demand. "High-end" cigar brands, which have larger profit margins and faster growth rates, now sell for between $10 and $20 apiece. Sales of premium cigars have shown a compounded annual growth rate of 35.6 percent, according to the chief financial officer of the largest cigar manufacturer. At this writing, the cigar boom shows no sign of slumping.

What price does a cigar smoker pay, beyond the actual cost? Are these particular nicotine-delivery devices as safe, nonaddicting, and charming as their proponents claim? Consider a few facts:

Cigars range in weight from about 1 g to 22 g (cigarette tobacco weighs less than 1 g). Their nicotine content ranges from 6 mg to more than 300 mg (cigarettes range from 6 to 11 mg, with the smoker obtaining 1 to 3 mg). It is not yet known exactly how much nicotine a cigar smoker typically obtains from each cigar. A cigar's nicotine concentration ranges from about 5 mg/g to 22 mg/g. In other words, there is little consistency among cigars. Their pH ranges from about 6 to more than 8 (cigarette pH is 5.5 to 6.5). This more alkaline product results in harsher smoke and a unique odor. Higher pH cigars can provide a more rapid nicotine delivery than other cigars that might have more nicotine. Size alone does not predict nicotine content, since nicotine content varies among cigars. Smoking style also varies, with some cigar smokers inhaling cigar smoke directly, others inhaling little directly but inadvertently inhaling environmental smoke, and some cigar users not lighting up at all but still absorbing nicotine from the unlit cigar.

Cigars differ from cigarettes in several other ways. A cigar is wrapped in tobacco leaf or in paper soaked in tobacco extract. A machine-made cigar could be produced in a process similar to that used for cigarettes. The larger, more expensive premium cigars are rolled by hand. The nicotine content of a cigar varies considerably from puff to puff, with nicotine delivered differently at the beginning of the cigar than at the end. The tobacco used in cigars is higher in nitrates than that used in cigarettes, resulting in greater concentrations of nitrosamines in inhaled and sidestream smoke. Although the risks of cigar use to the smoker are difficult to quantify because of the wide variation in usage, the environmental smoke from cigars is another matter. A cigar smoker produces about as much environmental or secondhand smoke as seven cigarette smokers, and the smoke from a cigar is believed to be more physically irritating to nonsmokers than cigarette smoke.

It can take a cigar smoker an hour to smoke one large cigar. As Henningfield explained in a 1996 report, "[T]he consumption of a few fat cigars could produce the daily burden of smoke exposure produced by the consumption of a pack of cigarettes." Henningfield and his associates reported in 1998 that the smoke acidity of larger cigars changes as they are smoked, which affects cigar users' ability to consume these cigars without experiencing nicotine-overdose effects.

Additionally, those who first smoked cigarettes and switched to cigars are more likely to inhale cigar smoke than are those who never smoked cigarettes. Therefore, the reduction in health risk by switching from cigarettes to cigars is likely not to be as great as the smoker might want to believe.

A 1996 U.S. national tobacco survey by the Robert Wood Johnson Foundation found that among students ages 14–19, an estimated six million had smoked a cigar during the preceding year. Of these, an estimated 1.7 million were female smokers. Having smoked a cigar increased by three times the likelihood of also smoking cigarettes. Among those 69 percent of students who did not report smoking cigarettes, males were more likely to have reported smoking at least one cigar. Students using smokeless tobacco were more than three times as likely as nonusers to have smoked cigars. Curiously, cigar smoking among these teenagers did not vary by locale, ethnicity, or race.

A 1996 Massachusetts Department of Health survey of students in grades six through twelve found that 10 percent of sixth graders had smoked a cigar; of high school students in grades nine through twelve, 28 percent reported smoking a cigar in the previous year, with half of those reporting smoking during the previous month. As with the Robert Wood Johnson study, a strong relationship emerged between cigar use and other tobacco use.

A third 1996 survey of ninth graders, this one conducted by the Roswell Park Cancer Institute in New York, corroborated the findings of the other surveys. Some 20 percent of boys and 6 percent of girls responding in Erie County reported having smoked a cigar in the previous month. In Chautauqua County, 24 percent of boys and 5 percent of girls also reported having smoked a cigar. As with the other studies, students who smoked cigarettes were more likely to report smoking a cigar than were nonsmoking students. Among those students using smokeless tobacco, about 63 percent also reported smoking at least one cigar during the previous month.

It's Not Your Father's Cigar Anymore

1992	year *Cigar Aficionado* magazine started publishing
100 million	total cigars sold in the U.S. in 1992
176 million	premium cigars imported into the U.S. in 1995
294 million	premium cigars imported into the U.S. in 1996
4.5 billion	total cigars consumed in the U.S. in 1996
700 million	projected annual premium cigar sales by the year 2000
6 million	teenagers who have smoked a cigar in the last year
22%	teenage boys who have smoked a cigar in the last month
11%	teenage girls who have smoked a cigar in the last month
14%	cigar-smoking teenagers who had never smoked before
up to 40 times more	nicotine in a cigar, compared with a cigarette
up to 40 times greater	maximum carcinogenicity of cigar tobacco, compared with cigarettes
famous cigar users	Demi Moore, David Letterman, Michael Jordan, Tom Cruise, Madonna, Wayne Gretzsky, Denzel Washington

These reports, published in the Massachusetts Medical Society's *Morbidity and Mortality Weekly Report* in 1997, were the first to estimate cigar use among U.S. young persons. Although the three surveys were consistent with each other, they did not include nonstudents of the same ages, and therefore they might not represent all adolescent cigar use.

Where are these young people buying cigars? They are buying them in most places where cigars are sold. Not only are cigars apparently readily available to youth who want them, but at this writing, they carry no warning labels. The Surgeon General's health warning that is legally required to be on many tobacco products does not cover cigars. Public health officials have expressed concern that many young users of cigars might not be aware of the health risks. The 1997 report citing the three surveys emphasized: "Immediate efforts should be made to publicize the health risks of cigar smoking; deglamorize the product in magazines, movies, and television programs; and protect nonsmokers from second-hand cigar smoke."

As for those risks . . .

The daily use of cigars increases the risk of oral cancer, lung cancer, chronic obstructive pulmonary disease, and heart disease, with risk related to the frequency of use and the degree of inhalation. Also, cigars contain nicotine, a substance with considerable potential for dependence. These risks were affirmed in a 1998 monograph released by the National Cancer Institute.

In apparent defiance or ignorance of the health risks, the fad is booming. For example, a video is now available to introduce women to the basics of selecting and enjoying "fine cigars." Marketed by an attorney in Seattle who with her husband owns a cigar store/smoking lounge, the video provides information about flavor, ring gauge, price, and brands, as well as cutting, lighting, holding, and smoking cigars. It also discusses wines, ales, and coffees that complement and enhance the cigar-smoking experience. It is a follow-up to a previous video aimed at educating cigar novices; unlike the first video, the second one is aimed specifically at women who want to smoke cigars.

Michael Erickson, director of the U.S. Centers for Disease Control and Prevention's Office on Smoking and Health, referred to the cigar craze as "a phenomenon of the 90s." The next few years will tell whether this fad has the momentum to build its own little bridge into the twenty-first century.

Donald Nelson, editor of the *Puget Sound Business Journal,* explained the cigar phenomenon this way: "Every generation celebrates its independence from its predecessors by lapsing into a kind of momentary lunacy that, on reflection years later, seems shallow, silly and pretentious.

"That's my take on the 20-something crowd's current wretched fascination with cigars. There's no other defensible explanation for the faux sophistication and phony elan that are part of the laughable cigar ritual.

"They're doing it to show us old geeks that they're the smart set, that they're really something." What Nelson said he "gets" is the rebellion part; what he doesn't "get" is the cigar part. "Cigarettes are intolerable, but cigars are—vile. They stink. They offend. And despite the hype about the exotic pleasure of a fine cigar (now there's an oxymoron!), a cigar is just another death-dealing tobacco stick that people have been hoodwinked into paying a lot of money for."

And yes, he conceded, his bias did show, for which he made no apology. And he did remember his own days of rebellion. He remembered wearing bell-bottom pants himself. "They were dumb back when I wore them, and they're still dumb."

So why do people smoke cigars? Evidently, the reasons people try cigars and keep using them are similar to reasons for cigarette use: marketing, nicotine, marketing, sensory experience, marketing, images of prosperity and well-being, marketing, keeping your mouth busy, marketing, being "in your face," marketing, relaxing with your friends, marketing—did we mention marketing?

CHAPTER 9

Like most smokers, she has tried to quit repeatedly, only to bump into the wall of her dependence on nicotine.

"If I don't have it, my personality changes, until I have completed withdrawal. I cry a lot," she explains. "Then once I'm done with the withdrawals and I just have to deal without having a cigarette in everyday life, I tend to get a little snappy."

She tends to be a little snappy sometimes anyway, she muses.

"The first few days after I quit, I cry—at nothing! I cry at nothing. I can be sitting here talking to you one second, and I'll burst into tears the next second, for no reason at all. There's no emotion—it's just a physical reaction. Boom, I'm crying, and I can't stop. There's no external force. I don't feel pain.

"I'll give you an interesting example of this happening. I quit for— it was my second day after quitting, and I went out with my friends to play darts. I was standing there playing darts, having a grand old time, and we were laughing and joking around. I wanted a cigarette, but, hey, this was my second day, and I was gonna be fine.

"It was my turn to throw the darts. I stepped up, I positioned myself, I went to throw the dart, and boom, I start sobbing. Uncontrollable sobbing. I couldn't stop. I *could not* stop. Boom. Done. It was the weirdest thing.

"Looking back on it, I know that it was just my body reacting to the nicotine withdrawal. So I wasn't feeling emotion, but I was really trying to curb that desire to smoke.

"My best friend came over. I kept sobbing and babbling for about ten minutes. Finally she said, 'Here, have a cigarette.' I just could not stop crying. That time, I'd quit cold turkey. The next time I decided I'd better use the patch."

So, what triggers a relapse, with or without the patch?

"I think I let down my guard. Usually I can have one drink and I feel that I'm okay, but the second that I have that second drink, if there's anyone around me smoking, then GOD! I WANT A CIGARETTE!

"I have to quit drinking to quit smoking. Even after six or seven or nine months, I'll decide one night that I want a glass of wine with dinner. Beer. Whatever. With friends. I don't go out on a drinking binge or anything. I just have a glass of something, and that's it.

"And if I have one drink and I manage to not smoke, then in a couple of weeks I do it again. Each time I do it, my resolve to not smoke gets less and less.

"And I always smoke more on opening night. Every chance I get, I smoke on opening night. Opening day. Opening day, if I'm at work during the day, I tend to take more breaks. Go outside more. As soon as I get to the theater, whenever I can get a break, I go outside and smoke.

"So, yeah, high stress will do it too."

Learning to Quit

They started coming in 1986, a group of health professionals visiting a small village in the southwest corner of Viti Levu, the main island of Fiji. They came only a few days at a time, yet the more than 200 villagers sensed the commitment of the medical team. They set up a dispensary and a system for safe water catchment. They helped eradicate scabies, trained village health care workers, and set up a local health care committee.

Then in 1990 they noticed that smoking had increased dramatically among the 238 Fijians in the village of Nabila. As tobacco advertisements had lured the Fijians into spending as much as a fourth of their small salaries on cigarettes, nearly half of all adult males in the village had become smokers. Almost a third of the entire adult population of the village smoked. Smoking had doubled in about five years. Health effects were becoming chronic: Hypertension was at an epidemic level. Asthma, vascular disease, and degenerative diseases were common.

The medical team had been advising villagers against smoking, handing out American Cancer Society posters and pamphlets, and warning villagers about the adverse health effects of smoking. The village minister set an example by abstaining from tobacco. It helped, but it wasn't enough. The visiting health team considered numerous strategies. What would work best in this population? Rapid inhalation? Social contracting? A reward system? The health team considered offering to build the village a community center if every smoker in the village would abstain. Team members were concerned, however, that they were imposing their own value system on another culture. Would this alienate youth from their elders? Would it undermine village authorities? How would it affect village smokers who later relapsed? The team expressed their concerns to the village spokesperson, who relayed them to the village elders.

The spokesperson made an announcement as the team's visit concluded: The entire village had decided to abstain from smoking. The idea had actually been proposed by the village youth. The money saved by not buying cigarettes would be contributed toward a community center. The medical team offered a donation toward the center; the village matched the funds. The elders declared that, soon, smoking would be forbidden by a formal tabu.

Was it too ambitious a plan? How could an entire village quit, when the chance of relapsing is 70 to 80 percent in well-designed community smoking cessation programs? The medical team departed, hoping for the best. Three months later, the village health committee sent the medical team a letter declaring that Nabila was now a nonsmoking village. The village had designed and implemented its own cessation program.

Before the formal tabu commenced, smokers collected all the cigarettes in the village. At a meeting, the villagers who used tobacco chain-smoked until they were nauseated. The remaining cigarettes were ceremonially destroyed. That night, the village commenced a six-hour ceremony in which they drank kava, a powdered root that is mildly relaxant, euphoriant, and hallucinogenic. Drinking kava is a sacred experience that can help create change or enforce a tabu. Most Fijians believe that breaking a tabu can result in bad health or injury.

You Quit, but It Doesn't

6 – 12 hours	how soon withdrawal symptoms start after the last use of tobacco
1 – 3 days	period in which withdrawal symptoms are the worst
3 – 4 weeks	how long withdrawal symptoms usually last
more than 40%	smokers whose withdrawal symptoms last longer than 3 to 4 weeks

(Hughes, Higgins, and Bickel, 1994.)

The "evil spirits" of the cigarettes were allowed to enter the kava that remained unused, and the kava was thrown to the ground. The villagers believed that this would help eliminate the desire to smoke. At this point, the tabu commenced.

The village's nonsmoking pledge was posted permanently. Fijian news media carried reports of "the village that quit smoking." Even some 50 Nabilans who smoked but no longer lived at Nabila honored the tabu by quitting smoking. Three-fourths of the young people in a neighboring village also quit smoking, in a gesture of solidarity.

Most of the ex-smoking villagers reported having little trouble quitting. Those who had problems quitting sucked on lollipops or reinforced their commitment through another kava ceremony. Four relapsing smokers did experience negative consequences. One tripped and cut himself, another was attacked by a dog, a third developed testicular swelling, and the fourth became unconscious after smoking during a kava ceremony. They all sought forgiveness and quit smoking.

Nine months after the tabu started, almost no one in the village smoked. About two years after the tabu, smoking rates remained very low. Four elderly persons and a youth visiting from another village were allowed to smoke.

The researchers reporting this account remarked about the medical team's joint trust, understanding, and commitment with the village. Australians Gary Groth-Marnat and Simon Leslie, and Mark Renneker of the University of California at San Francisco, also noted that the repeated visits of the medical team added the expectation of accountability to their relationship with the village. The nature of Fijian society and culture also enhanced the plan's possibility of success. Nabila was a cohesive community where Fijian beliefs in group harmony and respect for authority were beneficial. Additionally, the villagers expected their cessation attempt to be successful, in part because they believed in the power of the kava and the tabu. Also, the negative emotions, interpersonal conflicts, and social pressure often contributing to relapse in Western cultures were minimized in Fijian culture. The Fijians did not hesitate to consider cigarettes as evil and immoral, and thus the "powerful motivator of morality" helped reinforce their commitment to quitting.

The health team proposed several suggestions for health workers in similar situations:

- Indigenous people should develop their own programs.
- Consider unique rituals that could increase the power of smokers' decisions to change.
- Enhance change by working with healers or other persons of status in the community.
- Consider health promotion in relation to the culture's existing values.
- Do not expect or demand change too early.
- Gradually develop a committed relationship over time.

Fijians' communal decision making provided a contrast to Western society's focus on individualism. This is not how most people in Western cultures quit smoking. Attempts to develop a sense of community among quitting smokers in the West have not resulted in dramatic quit rates. However, successful Western community-based treatment programs have involved the community in making decisions and designing the program, have considered cultural issues, and have offered long-term help.

As the story of the Fijian village illustrates, the initial act of quitting smoking is not enough to ensure that a person will remain a nonsmoker. Equally important is the necessity of staying quit over the following months and years.

A brief return to smoking, involving perhaps one or two cigarettes, is termed a *lapse*. It is common for smokers attempting abstinence to indulge in a lapse episode. Often, however, a lapse leads to a full-blown *relapse*, in which the individual returns to ongoing smoking or other tobacco use. The nature, prediction, and prevention of relapse constitute an entire realm of study within the larger field of tobacco cessation research. Scientists know, with some imprecision, who is likely to relapse, and what can help prevent it. Researchers also know that it is important to distinguish between lapse and relapse, since they have different impacts on quitting.

At what point does a lapse episode become a relapse? A task force of the National Working Conference on Smoking Relapse deliberated this issue in an attempt to find sufficient unity to allow comparability of research studies, as reported by Sally A. Shumaker and Neil Grunberg. The task force determined that seven consecutive days of smoking at least one puff per day would constitute a relapse. Obviously, this definition does not imply a return to baseline smoking levels, but it does imply a return to regular, repeated smoking.

Consider these accounts of lapse and relapse:

An unmarried man in his twenties quits smoking, with the help of a cessation treatment program. His co-workers, friends, and family are supportive and encouraging. One night, he goes to a bar. Although most others there smoke, he refrains. He reunites with a woman he has known intimately for some time. Late that night, they go to her home, where they spend the night. Early the next morning he wakes up, rolls over in this familiar setting, takes a cigarette out of the woman's cigarette pack on the nightstand, and smokes it. A few minutes later, as the fog clears from his mind, he remembers that he is a nonsmoker.

Or, rather, he remembers that he *was* a nonsmoker. Now that he has lapsed with this single cigarette, he is no longer sure whether he should consider himself a nonsmoker. At work that morning, he calls the clinician who has guided his stop-smoking group through cessation. He tells her of the cigarette he unthinkingly smoked. She assures him that it is only a lapse episode and does not need to lead to a full-blown relapse. He can continue his smoking cessation from this point.

Another situation: An engaged college-age couple decide to quit smoking together. It becomes a cornerstone of their commitment to each other. They plan their wedding and determine that they will honeymoon in their favorite remote, mountainous area. They decide that they will use their honeymoon as a time to quit smoking, together. They exchange vows and head off with their backpacks, throwing away their cigarettes as they embark on their backwoods honeymoon.

For a week, the hills are alive with the sound of their bickering. For the first time in their relationship, they fight and snip, spewing nicotine-free venom from the depths of their abstinence. It is no surprise that on their return to civilization, they also return to smoking.

And a third story: A psychology graduate student is enlisted to help lead a stop-smoking group. No one at his university knows that he secretly smokes. He learns the skills of helping smokers quit and becomes an effective interventionist. He helps dozens of smokers quit and stay quit; he learns every relapse prevention strategy, every tool of aversive conditioning, every problem-solving routine. He knows how to do it all for others but cannot do it for himself. He uses the nicotine patch and nicotine gum to get through times he cannot smoke openly.

Years pass; he graduates and does postdoctoral work. Finally, he quits using nicotine completely when he becomes a parent. He would never smoke around the baby, he explains.

Every ex-smoker's course through withdrawal and into abstinence follows a unique path. Not only do withdrawal symptoms vary greatly from ex-smoker to ex-smoker, but the risks for relapse are also individual. What tempts one smoker might not tempt another. Each smoker's capacity to cope with temptation, lapse, and relapse depends on a host of somewhat unpredictable factors.

"What worked for me was . . ." So the story starts, whenever we ask a long-time ex-smoker how he or she quit and stayed quit. The methods are so various as to be uncountable and difficult to catalogue. Even if we ask a current smoker who has tried to quit but failed, the wording isn't so different. "What worked the best for me was . . ."

Why Ex-Smokers Relapse

People who haven't "been there" can find it hard to understand why anyone would relapse, once a smoker gives up tobacco and gets past the withdrawal phase. If the abstinence symptoms are history, why is it so hard? The answer can be explained partly by a description of how nicotine works. Nicotine is a reinforcing substance, which means that using it results in sensations and conditions that are perceived as positive by the tobacco user. It can help a smoker regulate his or her mood. It can help curb appetite and can help keep body weight at least a few pounds lower. It can heighten thinking and reasoning skills, although not dramatically. Some people find that nicotine enhances memory, eases anxiety and tension, makes sensory experiences feel more intense, and makes pain easier to bear. Not all nicotine users report all these occurrences.

Since these effects are pleasurable, they are reinforcing. This means that they become psychologically and mentally linked to the act and circumstances of smoking. These linkages of what psychologists call *stimulus* and *response* operate on the same principles that led Pavlov's dogs to salivate when they heard sounds associated with being fed. They are the same processes that start your mouth watering when you see a picture of your favorite chocolate. A sequence of behavioral psychologists, including B. F. Skinner and Albert Bandura, have described these phenomena well and expanded on their meaning in our social lives. If anything in psychology has been demonstrated to the point where theory has become doctrine, it is these principles.

When we use tobacco in any setting at all, what we perceive as the positive effects of using the substance reinforces our use of it. Just as we shape a child's behavior with reinforcement ("That's a magnificent mud pie, and thank you for not bringing it in the house") we also shape our own tobacco-using behavior by our very use of tobacco. In a laboratory, a researcher can teach a pigeon to peck in a certain spot on a cage wall by reinforcing the bird's behavior with food as it pecks closer and closer to the designated spot. An animal trainer can teach a cat to dance, can teach an elephant to stand on two legs, or can train a dog to jump through hoops, by using the same principles of reinforcement. We likewise train ourselves to reach certain emotional and mental states through the reinforcing use of tobacco. This results in a different, and perhaps more pernicious, form of dependence than what we would commonly call "addiction" to nicotine. We learn not only to depend on nicotine, but also to look to tobacco for its reinforcing effects.

The effects of reinforcement are multifaceted. When we smoke in a social setting, we are not only continuing our dependence on nicotine and experiencing tobacco and nicotine's many physiological properties, but we also are linking the experience of smoking to the setting. Tobacco use becomes tied to chatting with friends, sharing conversation, flirting, solidifying relationships, or whatever else. We engage in much more than just a physically reinforced action; we also reinforce our smoking or other tobacco use with the effects of the setting in which we use tobacco.

In addition, repeated exposure to nicotine leads to a physical dependence, such that the nicotine-dependent person requires nicotine to avoid experiencing adverse effects. A dependent smoker who does not get the dose of nicotine that the body expects and needs will begin experiencing withdrawal symptoms. The symptoms vary considerably between individuals, but they generally involve some constellation of the following effects:

- Anxiety
- Irritability
- Difficulty concentrating
- Restlessness
- Impatience
- Hunger
- Tremor
- Racing heart
- Sweating
- Dizziness
- Nicotine craving
- Insomnia or other sleep disturbance
- Headaches
- Digestive disturbances
- Depression.

Not every nicotine-dependent person will get all the symptoms, or will get them all at once. The symptoms may change throughout the course of withdrawal. A nicotine-dependent person using nicotine replacement (e.g., the patch or the gum) to help quit using tobacco may experience symptoms to a lesser degree, but may still experience them somewhat. Some clinicians have reported success with enhanced doses of nicotine replacement, through application of multiple patches or combinations of gum and patch. A smoker in withdrawal will begin to feel the symptoms fading within minutes after he or she uses nicotine. This is evident from subjective self-reports as well as from computerized testing administered to smokers deprived of nicotine.

Not only will the nicotine-deprived smoker undergo withdrawal symptoms during abstinence, but he or she will also notice the absence of nicotine's purportedly enhancing qualities. For instance, the mind may feel dulled because of the effects of withdrawal, but also because of the absence of the nicotine that provided the mild enhancement. In addition, the lack of nicotine may affect how medications and other substances are taken into the body, perhaps resulting in symptoms secondary to withdrawal but noxious nonetheless.

The First Month after Quitting

adrenaline	decreases
heart rate	decreases
thyroid function	decreases
tremor	decreases
vigilance	decreases
aggression	increases
anxiety	increases
caloric intake	increases
craving for cigarettes	increases
depression	increases
difficulty concentrating	increases
hunger	increases
irritability	increases
metabolism of some drugs	increases
nocturnal awakening	increases
resting metabolic rate	increases
restlessness	increases
sweet/fat food intake	increases
taste for sweets	increases
weight	increases

The experience of going without nicotine, resulting in withdrawal symptoms and other unpleasantness directly or indirectly caused by the absence of nicotine, can be a powerful precursor of relapse. Withdrawal, however, is over within a few weeks of cessation. Why, then, do many ex-smokers relapse long after that?

An answer to that question lies in the behaviors surrounding tobacco use. The modern cigarette is a particularly effective means for providing rapid reinforcement, which sets it up as a coping response in many situations. Inhaled nicotine, delivered through tobacco smoke, is rapidly absorbed by the body. About the time the cigarette is snuffed out, blood levels of nicotine have peaked. Behavioral scientists have long known that immediate and rapid reinforcement can create a powerful link between situations and behaviors. If the reinforcement were to come hours or even minutes later, the sensations and events would not be tied so tightly to the use of tobacco. Because smoked tobacco is rapidly reinforcing, the reinforcement from nicotine in cigarettes becomes a powerful mechanism, *independent of its addictive potential.*

Smokers can come to feel dependent on nicotine to help them maintain their normal state of functioning. A tobacco user who has relied on nicotine throughout many times of high stress may come to believe that he cannot manage stressful events without nicotine. Someone who uses nicotine in conjunction with alcohol in social settings may feel inadequate without a cigarette, because he is used to the way nicotine helps him relieve anxiety, or because he is used to the combined interaction effects of nicotine and alcohol.

Such highly reinforced use of tobacco creates a high potential for relapse in people who use it to maintain a normal state. The nicotine delivery system we call a cigarette affords the smoker a range of emotional and mental effects. A smoker quickly learns to vary the rate and intensity of smoking to maximize his or her preferred effects.

Prior to relapsing, many ex-smokers think that they are in control of their smoking behavior. When they slip, they may feel aspects of self-blame, guilt, self-criticism, depression, and hopelessness. Rather than pushing them back toward abstinence, these feelings actually increase

Missed Opportunities

70%	smokers who see a physician each year in the United States
more than 50%	smokers who see a dentist each year in the United States
2 million	estimated additional smokers who would quit annually if 100,000 health care providers helped 10% of their patients quit smoking
11%	estimated U.S. health insurance carriers that provide coverage for treatment of nicotine dependence

the likelihood that the ex-smoker will relapse. The complex cognitive process involved in this unfortunate twist of direction has been called the *abstinence violation effect*. According to behavioral scientist George Marlatt, a high-risk situation that may result in relapse includes these three aspects: (1) the expectation of the positive effects from a lapse, (2) the actual reinforcement from the lapse, and (3) the pressure pushing the ex-smoker toward lapsing. In other words, an ex-smoking woman facing a sudden, unexpected stress might think that a cigarette surely would help her cope. Then the cigarette *does* help her feel as if she's coping. In a situation where smoking would be tolerated or even encouraged, she could find herself lapsing, perhaps lapsing again, and eventually relapsing. She might then feel negative about her inability to keep from smoking. Feeling depressed and discouraged, she might not find it easy to return to abstinence. Instead, she might return to her pre-cessation level of smoking. And if she feels bad about *that*, another cigarette surely would help her cope.

Looking into the Statistical Ball

There's no magical crystal ball that enables us to see who will quit and stay quit. There are, however, some common traits among those who are successful, as well as among those who fail on a given quit attempt. Also, some life situations, traits, and behaviors seem to predispose people not only to smoking, but also to quitting and to staying quit.

Just how powerful a preventive or therapeutic intervention can be depends on many factors. The degree of dependence on nicotine can influence the success of quitting and the likelihood of relapse. A smoker's tobacco-related health problems can provide motivation to quit, but can also complicate quitting. The smoker's lifestyle has a profound effect on the possibility of quitting successfully. A person with supportive family and friends, nonsmoking home and work environment, low social stressors, and a stable life is likely to find it easier to stay quit than someone who is immersed in financial problems, strained relationships, stressful situations, and a social environment where smoking is tolerated or encouraged.

Who would be the ideal candidate to stop smoking for good? He (the person would be male, according to the statistics) would be a light smoker who hasn't smoked long, who does not have a history of failed quit attempts, who is not particularly overweight, who is willing to comply with treatment instructions, and who is capable of learning and using coping strategies and techniques. Swedish researchers Per Tillgren and colleagues found that the best predictors of success among nearly 13,000 Swedes who quit smoking were lack of quitting attempts during the previous year, participating of one's own volition rather than being recruited by a nonsmoker, and being married or cohabiting. If we consider the inverse of most of these criteria, we are envisioning someone who would be likely to have far more trouble quitting. Heavy smokers, women, overweight smokers concerned about gaining weight after quitting, and people unwilling or unable to employ coping techniques are less likely to succeed.

Men and women experience smoking and quitting differently. Overall, women appear to be less responsive to nicotine replacement as part of smoking cessation. If this finding is supported by further research into cessation for women, it is possible that sex-specific treatment approaches could be developed. It may be that women will benefit less from nicotine replacement, but will have greater success by learning to cope with the nonnicotine reinforcement of smoking, such as the sight, smell, and taste of tobacco. Even so, Ken Perkins emphasized in 1997 that women do experience nicotine withdrawal, which can be relieved to some extent by nicotine replacement. However they may benefit from much more than just replacing nicotine.

For both men and women ex-smokers, the most common triggers to relapse center around negative emotions and abstinence symptoms. Gary Swan and colleagues at SRI International related in 1996 that ex-smokers who reported heightened anger, depressed mood, and craving for cigarettes were more likely to relapse quickly. When other quitting-related factors were taken into consideration, the prominence of craving for cigarettes still significantly predicted a higher rate of relapse among ex-smokers.

A smoker's beliefs as he or she considers quitting can markedly affect the chances of success. A research team from American University in Washington, D.C., interviewed 100 people who had recently quit smoking, asking them why they believed they would or would not be able to stay quit for a year, and what benefits they expected. David A. F. Haaga and his colleagues then coded the responses according to dimensions of belief about success in abstinence, including health factors, the experience of quitting, external factors that could influence staying quit, the role of nicotine addiction, and personal attributes. Subjects' responses regarding the benefits they expected from quitting included reduction in inconvenience, lower expense, reduction of health problems, diminished health risk for themselves or someone else, improved health, and improved feelings of psychological and physical well-being.

The most commonly cited reason for the expectation of continued abstinence was the belief that the ex-smokers' own personal traits would determine their success. About half of the people who gave this response mentioned the concept of willpower or self-discipline. Few of them thought that specific actions or tactics would help prevent relapse. Improved physical health was noted most often as an expected benefit.

Worse Than Hard Drugs?

It seems that many smokers who have difficulty quitting console themselves with the knowledge that the abstinence symptoms experienced when quitting smoking are purportedly worse than those associated with quitting other addictive substances. This sentiment usually is expressed in a phrase such as, "I've heard that addiction to nicotine is the worst addiction you can have, even worse than heroin or alcohol." The unasked question ("Is this true?") can be approached from various angles. Exactly what makes dependence on a substance bad or makes it worse than dependence on another substance is a matter of judgment. Some might give more weight to the immediacy of the physical threat or to the nature of physical or cognitive impairment in determining the severity of an addiction. It is also possible that some judge the relative badness of a substance by the unpleasantness of the symptoms that arise when one is abstinent.

The most direct scientific comparisons between nicotine and other substances of abuse involve analyses of abstinence symptoms. British researchers Robert West and Michael Gossop in 1994 described a shift in emphasis from physical withdrawal syndromes to compulsive use, in efforts to define the concept of addiction. They outlined the basic difficulty scientists encounter in studying abstinence effects, which mirrors the confusion smokers may experience as they attempt to quit. Scientists' prior belief that physical withdrawal was "the defining feature of drug addiction" has matured into viewing addiction more as compulsive use. The features of withdrawal syndromes vary across classes of drugs, as well as across individual experiences.

John Hughes and colleagues of the University of Vermont's Human Behavioral Pharmacology Laboratory concluded in 1994 that nicotine has several commonalities with other substances of abuse, but the similarities and differences are difficult to make sense of because so much about nicotine withdrawal remains untested. They reviewed dozens of pertinent studies and noted these comparisons:

- Three signs and symptoms distinguish nicotine abstinence effects from those of sedatives and opioids: decreased heart rate, increased eating and weight, and absence of observable physical effects.
- The mood disturbance common in nicotine abstinence is similar to that treated at outpatient psychiatric clinics.
- Nicotine withdrawal is less severe than withdrawal from alcohol and opioids, and it does not result in significant medical or psychiatric difficulty. However, some smokers experience more severe abstinence effects related to nicotine dependence than addicts experience when they quit using sedatives or opioids.
- The onset of nicotine abstinence effects occurs sooner than those for alcohol, caffeine, or cocaine, but at about the same as heroin's.
- Nicotine abstinence effects peak later than caffeine withdrawal, but peak on about the same time frame as withdrawal from alcohol and heroin.
- Withdrawal from alcohol, opioids, and stimulants occurs over an extended time period during which symptoms persist. Except for hunger increase and cigarette cravings, which persist for at least six months, nicotine withdrawal does not have protracted symptoms.
- Withdrawal from some substances, such as alcohol, involves a series of stages. The early tremor, sweating, and insomnia of alcohol withdrawal are later followed by seizures, disorientation, and hallucinations. Parallel stages have not been described in nicotine abstinence.

Is the possibility of severe abstinence effects part of what keeps someone using an addictive substance, and part of what leads to relapse in former users? Although these ideas make intuitive sense, the data do not always follow. The link between abstinence effects and relapse remains tenuous, awaiting more solid scientific replication and confirmation.

Studying abstinence effects and the potential for relapse is a venture into complexity. A basic limitation of this area of research is the problem of generalizability. People who volunteer to be part of nicotine cessation research, or who attend clinics for help in quitting, might not be representative of tobacco users as a group. Only a small minority of smokers seek formal help for quitting smoking. Of those smokers, not all will join or stay with a cessation program. Additionally, the process of being studied during cessation may affect the outcome in ways that are hard to predict or assess.

A related problem is the nature of withdrawal itself. The above-cited list contains many symptoms that are difficult to quantify, and that can also be influenced by other life factors. For example, one typical abstinence symptom is irritability. Many things can make us irritable; going through nicotine abstinence may be only part of the reason for increased irritability. Even more complex is the symptom of depression. A smoker who may be using nicotine to help curb the feelings associated with depression may find those symptoms worsening during depression. But is this due solely to the sudden lack of nicotine in the system? If nicotine has merely been helping mask feelings of depression, is nicotine abstinence actually increasing depression, or is it merely unmasking depression? Or is the equation even more complicated than that: Is depression at least in part the result of the social and emotional consequences of giving up a substance of dependence? Only the most cautious and careful research can tease apart such questions.

The best way that scientists work around the thorny issues of assessing withdrawal symptoms is by doing what West and Gossop described, which involves defining "a specific set of criteria for what counts as presence of withdrawal using a specified measure." Additionally, researchers have to "avoid falling into the trap" of overinterpreting their findings beyond the scope of their definition of a withdrawal symptom. Which is to say, if they are studying the emergence of anxiety symptoms in withdrawal, they must be careful not to draw conclusions about other aspects of anxiety beyond those symptoms.

What does this matter to the average smoker who wants to stop? It may not seem to matter much to someone who throws away his or her cigarettes one day and grits through withdrawal by raw force of will, regardless of the unpleasantness of the experience. It may matter greatly, however, to an ex-smoker who recently quit and who now feels irritable, hungry, and depressed, and who wants to know how long these feelings will last. It also matters if science hopes to alleviate or ameliorate withdrawal symptoms, or to develop strategies to help people avoid relapse.

Curiously, the nature of abstinence effects may affect the success of a quit attempt in unpredictable ways. Research indicates that those who are more dependent on nicotine have an increased tendency to relapse. However, sometimes severe abstinence symptoms become a deterrent to use of the substance itself. Also, severe symptoms that occur soon after cessation can heighten resiliency; a smoker who has had an easy time for the first couple of weeks can be blindsided by the sudden worsening of such symptoms as cigarette cravings or hunger that may emerge as abstinence progresses.

In any case, it is clear that the experience of withdrawal differs from individual to individual, and that some people have a higher tolerance for discomfort than others. Also deeply intertwined with the issues of cessation and relapse is the smoker's milieu. As University of Michigan researcher Ovide F. Pomerleau noted in 1992, "Smoking is becoming the habit of the disadvantaged and the less affluent and less educated." This change in the worldwide demographics of smoking has ominous implications for success in eventually quitting. Those with limited knowledge of the health risks of smoking are also those with reduced access to programs, medications, and personnel who could help them quit. A middle-class working woman in the United States who smokes can, if she chooses, go to virtually any drug store, discount department store, or grocery and spend a relatively small portion of her take-home salary for nicotine replacement medication. She can see her physician and obtain a prescription for a stop-smoking pharmaceutical product not yet available over-the-counter, usually with her physician's encouragement and blessing, and sometimes at the expense of her health insurance company. She can sign up for a stop-smoking class through a hospital or a local organization that sponsors stop-smoking groups. She has many options, all of which increase her chances of quitting and staying quit.

A herdsman on the Bolivian altiplano does not have these options. A Fiji islander generally does not. A smoker in any developing country, or in less urban parts of a developed country, does not. Those with poverty-level incomes do not. Even affluent U.S. teenagers who hide their smoking from their parents do not. Science has already found many tools for helping smokers stop smoking, but science cannot circumvent the social, economic, and political barricades that keep these treatments from being widely available. Science can address the issues within its domain, but science has yet to find ways to prevent tobacco use from being widespread among those who have the least access to helpful interventions or to adequate health care.

A Case of Optimism

The British Royal College of Physicians and the U.S. Department of Health and Human Services have both made it clear: Cigarette smoking is their countries' primary self-imposed health risk. In light of this fact, many researchers have pondered why many smokers do not choose to quit. A common response to this question has been that, to some degree, smokers are irrational and have lost control over their behavior. It has even been suggested that smokers have "cognitive deficits," a polite term for a malfunctioning brain.

A more benign response to continued smoking has been to assume that smokers simply aren't aware of the risks. This has led to widespread public health campaigns in some developed countries, with smokers being bombarded with warnings about the dangers of their smoking. Public information blitzes have helped lower smoking rates in the United

States since the 1960s, but they have not provided a complete solution to the problem of continued smoking. Millions of smokers who know that tobacco is dangerous continue to smoke.

F. P. McKenna and colleagues of the University of Reading in England put a fresher face on this question with their 1993 report about smokers and optimism. After extensively questioning 120 smokers about health risks and expectations, they determined that smokers have what they termed an *optimism bias*. In other words, smokers rated their chances of developing tobacco-related health problems lower than did nonsmokers. Strikingly, however, both smokers and nonsmokers were equal in rating the health risks of the *average* smoker and nonsmoker. Apparently, a smokers' optimism bias applies only to himself or herself.

"There was no evidence of defensive denial in smokers about the likelihood of ill-health occurring to the average smoker," the authors wrote. "However, clear evidence of an optimism bias is present in that smokers clearly consider they personally are less likely to develop smoking-related diseases, compared with the average smoker." The illusion has its limits, however. Smokers did perceive that their health risk was greater than that of the average nonsmoker.

The authors concluded: "[A]lthough for smokers the illusion is present and powerful it is constrained suggesting that illusory optimism operates within specific boundaries." Those boundaries apparently extend to the smokers' friends and family as well. A 1998 study by John Pennington and James Tate of Middle Tennessee State University found that smokers' optimism bias extends to those close to them. Although the unrealistic optimism was substantial, it was confined to smoking-related disease.

CHAPTER 10

So, I ask her, how many times have you quit?

"I couldn't even tell you." She shakes her head and shrugs. "Dozen? I've quit for nine months twice, six months two or three times." She speaks with an air of confession. "These are times I've gotten past the physical addiction. Anywhere from three to nine months. Three months, many times. I can't even tell you how many times I've quit for two or three months.

"I've quit, oh, probably 16, 20 times. I quit every night."

And do you learn something each time? I expect her to explain what triggers her into relapsing, or what temptations are just too intense to withstand.

"Yeah, actually I've learned that autumn is the best time for me to quit. If I quit right around Thanksgiving, I tend to stay quit longer. I don't know why. Couldn't tell you. Thanksgiving, Christmas—maybe it's the change of season, change of temperature.

"I never have wanted to smoke. I've never pictured myself smoking ten years from now. Getting to it is difficult. I keep trying! I keep trying!"

When I ask what methods she's tried, she laughs.

"I don't consider any of them successful, because obviously I still smoke. The times that I quit for nine months, it's like a paradigm shift. You kinda have to brainwash yourself into the fact that you don't smoke. Like, if you're out and somebody says, 'Do you want a cigarette?' you say, 'No thank you, I don't smoke.' Or if somebody asks you for a cigarette, you say, 'No, *I don't smoke.*' You don't smoke. You tell yourself, 'I'm not a smoker. I don't smoke.' And then when that settles in, then you're not a smoker. You've gotta brainwash yourself that you are not a smoker."

So part of it is redefining yourself? But before she can answer, I interrupt myself with another notion: Maybe all of your quit attempts have just been rehearsals, and you're still waiting to quit for real?

"I know I just can't have that first one. I know the second I break down and have it, that's it! I've lost! The thing is that I can't drink," she explains. "When I quit smoking, I can't drink. When I drink, I want a cigarette. My craving increases. When I have hard alcohol, I don't have as much of a craving as if I have beer or wine. Different alcohols affect me differently. I get different reactions to them."

And which quit attempt was the most successful?

"I'd been using a half a pack a day—no, not even that much. A quarter of a pack. Five. And my best friend didn't smoke. She didn't like it. She talked me into quitting, so I quit. I stayed quit for nine months. Clove cigarettes were the big things. Nine months after I quit, she started smoking clove cigarettes. She'd have one every few days, so I decided to try one. Pretty soon I was smoking five clove cigarettes a day. I switched back to cigarettes. And then I was smoking clove cigarettes *and* regular cigarettes. Soon, I was up to a pack a day."

Her highest rate of cigarette consumption occurred over a five-month period in college, when she found herself surrounded by smokers

228

and smoking two packs a day. "I hadn't realized it. As soon as I realized it, I cut back down."

"Everybody smoked in the Theater Department," she recalls. "Things are getting better now working in theater, since people are quitting smoking. But it's strange to be working with an actor who's out there belting away this incredible song with a beautiful voice, great set of lungs—then he goes outside on his break and he's having a cigarette."

Smoking will always surround her, she knows.

"I'll just have to deal with it. Have to be strong," she declares. The theater will be a hard place to avoid smoking, "just because there's a huge incentive to smoke. More there because it keeps me awake, and focused. I read about the studies that say that nicotine increases your focus. And it works! It works great! When you're really tired, and you're getting spacey, and you know that you need sleep, but you can't, it works great. I'll just have to find other ways.

"I haven't quite figured out all the bugs yet."

As hard as she has found it, she has quit smoking during the production of a play. Sometimes, she says, quitting isn't that hard. Once she sought help in stopping smoking through a health maintenance organization. "They make you go to this class and do all this stuff—it's great! Stupid little class works great!" That time, she didn't smoke for three months.

She stops to calculate. The diamond solitaire on her left ring finger is a continual reminder: This year, she and her fiancé will be married. Soon after that, they hope, they'll start a family. They have decided that their children will not have parents who smoke. On this point they are united and firm.

No matter the cost in discomfort and aggravation, her days as a smoker are numbered.

Gimmicks, Gizmos, and Grit

For $95, a mail-order company in Waco, Texas, will send you Mr. Gross Mouth, a three-times-life-size hinged model of a mouth plagued by residual effects of using snuff tobacco. It is designed, apparently, to both deter potential tobacco users and terrify current users. If Mr. Gross Mouth's gingivitis and oral carcinoma aren't convincing enough, there's an even bigger model—Giant Mr. Gross Mouth. Weighing in at $135 and 12 inches wide open, Giant Mr. Gross Mouth has a four-inch-wide "cancerous" tongue made of realistic Biolike material.

Not to mention Mr. Dip Lip, whose flesh-like lips retract to show stained, deteriorated teeth and gums resembling those of smokeless tobacco users. Also made from Biolike is the Itty Bitty Smoker, a model of a ten-week-old fetus smoking a cigarette, "a hard-hitting reminder that pregnant mothers have special responsibilities," the company asserts. These educational products and an array of models of diseased lungs, folding displays, and "durable polyester" body-part models "painted in gory, realistic detail" are focused on developing an aversion to tobacco use, often among those who already use it.

The stop-smoking marketplace has no end of gimmicks and gizmos purported to help curb tobacco use. Among the most interesting are these:

- One cigarette-size device is touted in mail-order catalogs as "an absorbent cartridge that reduces up to 75 percent of the nicotine without affecting the cigarette's taste. Your craving is reduced, so it's easier to quit."
- Another device punches needle-size holes in cigarettes to reduce the inhaled smoke and thus help a smoker cut down on tobacco smoke exposure without having to cut down on cigarettes.
- A wristwatch look-alike device delivers post-hypnotic cues that reinforce suggestions from a daily compact-disk-based hypnosis session. The cue device works through visual, auditory, and tactile stimulation. It even has a panic button that can be pressed when the urge to smoke becomes overwhelming. It sells for $265.
- And then there's the lettuce cigarette. Leaves from romaine and iceberg lettuce are transformed into sheets, treated with enzymes, shredded, seasoned with herbs, and processed into a non-tobacco, non-nicotine cigarette. The device's inventor, a pharmacist, is marketing the lettuce cigarette as a stop-smoking alternative to tobacco. He anticipates that people will switch to it to break their addiction to nicotine while maintaining the rituals and experience of smoking. Eventually, they will wean themselves from the nonaddicting lettuce cigarette as well.

Nor is the attack on smoking limited to gadgets. A small notice in a local newspaper advertises a hypnosis clinic, where your smoking can be cured for $40. Because the hypnosis clinic will be conducted at a local hospital, the notice carries the imprimatur of the medical facility. But will it work? The same question comes up every time you see an ad for losing weight quickly and permanently: If it's so easy to lose weight, why are there so many overweight people? And if stopping smoking is as easy as being hypnotized, why do so many people still smoke?

To Quit or Not to Quit

40%	current smokers not considering quitting in the foreseeable future
40%	current smokers ambivalent about quitting
20%	current smokers intending to quit in the next few months
50%	nicotine-dependent adults who attempt to quit
more than 90%	quit attempts made without formal treatment
33%	those attempting to quit who are abstinent for at least 2 days
2.5–5%	quitters who stay quit for 1 year

Other quitter-hopefuls take a more circuitous route to quitting, by cutting back. Some people switch brands as they prepare for a pending quit date, going from a higher-nicotine to a lower-nicotine cigarette. This process is called *nicotine fading*. It has been used successfully in conjunction with several other stop-smoking methods. This technique is not applicable for smokers already using low-nicotine brands. Some smokers find smoking reduced-nicotine cigarettes to be aversive in and of itself. Also, smokers switching to a lower-nicotine brand may be taking in about the same amount of nicotine anyway if they compensate by smoking more cigarettes or inhaling and puffing differently.

Some smokers attempt to quit by switching to another form of tobacco, such as oral snuff, cigars, or pipes. At least one book and several scientific publications have promoted the use of smokeless tobacco as a supposedly safer alternative to smoking, which is satisfactory only for those not worried about the numerous health risks that smokeless tobacco entails. Other smokers switch to cigars, assuming that because they inhale cigar smoke less than cigarette smoke, and because they smoke cigars less frequently than they would smoke cigarettes, cigars are safer. Both of these switching techniques carry their own risks, which are substantial.

Techniques termed *aversion strategies* are also in the stop-smoking armamentarium. These approaches, based on what psychologists term a "behavioral" model, typically involve developing an aversion to cigarette smoke. The aversion technique that has been studied the most is rapid smoking, in which the smoker puffs every six to eight seconds until puffing is no longer bearable. This can be done several times at a series of sessions. Another technique involves doubling or tripling the usual smoking rate for several days prior to quitting, to achieve satiation in the smoker's home environment. Clinicians recommending these techniques must be careful to avoid introducing additional cardiovascular risk in patients. A low-risk aversive technique involves the smoker's saving cigarette butts in a sealed jar before quitting; when the temptation to smoke strikes, one sniff in the jar may be enough to deflect a potential lapse back to smoking.

Currently at the head of the class among smoking cessation aids are several pharmacologic treatments, including a nicotine replacement delivery device called "the patch," a small, adhesive bandage-like system that provides a slowly delivered, steady amount of nicotine to replace the nicotine that otherwise would be acquired from smoking cigarettes. Preceding "the patch" was "the gum," a vaguely neutral-tasting substance that delivers a fixed amount of nicotine if it is chewed and "parked" in the mouth correctly, which can be used to curb cravings and urges to smoke during cessation. Other nicotine delivery devices include a nicotine nasal inhaler, which has been tested both in Europe and in the United States. Nicotine replacement was designed to be a short-term means of weaning oneself off tobacco while dealing with the behavioral changes necessary for quitting smoking. However, long-term use can have its own negative consequences and thus is not recommended.

Each nicotine delivery approach has its own virtues. Some smokers prefer using nicotine gum because it allows them to regulate their nicotine replacement dose. Others prefer the patch because of its ease of use. The inhaler is preferred by some because it delivers nicotine via puffing, and thus shares some of the sensory characteristics of smoking.

Another pharmacologic smoking cessation aid is an antidepressant medication, bupropion, marketed as Zyban. In May 1997, this drug was approved by the U.S. Food and Drug Administration. A non-nicotine medication, it went through extensive clinical trials, as did all the nicotine replacement devices and the gum, before it was approved for use as a prescription medication.

Since the late 1980s, the same drug was marketed as the antidepressant Wellbutrin, or Wellbutrin SR. A scientist monitoring depressed patients who were using Wellbutrin noticed that some patients quit smoking. A multi-site trial of bupropion reported by Richard D. Hurt and colleagues in 1997 examined effects of three bupropion dosage levels and a placebo in more than 600 subjects who were quitting smoking. Smoking was reduced significantly in the groups of subjects who were given either of the two highest doses of bupropion, but not in those given the lowest dose or the placebo. Those subjects receiving the highest dosages of the medication also gained the least weight as they quit smoking. Even with that success, many of the participants in the study were smoking one year later. This medication currently is being evaluated further in multiple clinical studies.

The Zyban formulation of bupropion differs from Wellbutrin SR in two significant ways. The upper limit of use of bupropion for smoking cessation treatment is 300 milligrams; for depression treatment, it is 450 milligrams. Smokers are encouraged to use Zyban for cessation rather than Wellbutrin SR, because the higher bupropion dosage of Wellbutrin SR could unnecessarily increase the risk of seizures.

Ironically, the pharmacologic interventions, including those containing nicotine, must go through extensive testing for both safety and efficacy before they can be marketed, while tobacco remains a readily available and relatively inexpensive source of nicotine. In fall 1997, an international panel of tobacco experts urged governments throughout the world to ease restrictions on nicotine replacement as a way to help millions of smokers quit. As David Sweanor, a legal counsel for Canada's Non-Smokers' Rights Association, was quoted by Reuters news service, most of the world's 1.1 billion smokers want to quit but are addicted to nicotine. Quitting "is very difficult for them . . . because regulation of nicotine substitutes is far tougher than for cigarettes. The whole thing is upside down and it has to be corrected." In some countries, nicotine replacement requires a doctor's prescription. Because the limitations make the market small, the cost remains high. Smokers in developing countries thus have virtually no access to nicotine replacement.

A necessary part of successful tobacco cessation is modification of behavior. A small percentage of tobacco users will seek help through structured stop-smoking programs. The majority of those attempting to quit will try to do it by themselves, without help. To smokers, particularly the do-it-yourselfers, the array of devices and approaches can be bewildering. Many who have quit repeatedly, only to resume repeatedly, know what has worked for them, at least temporarily. They also know that tobacco is difficult to leave behind.

Cold Turkey on Wry

Some smokers believe that the best way to quit smoking is stop abruptly, or (as North Americans phrase it) to quit "cold turkey." They insist on gritting their teeth and toughing it out. This approach does work for some people. Similarly, it is a fact that most ex-smokers stopped without any formal help, although it typically takes many attempts before they finally succeed. In any case, the statistics predicting success for any given quit attempt aren't on the side of the cold-turkey quitter. Studies with thousands of smokers have shown repeatedly that the best way to quit smoking is to have help, and that cessation rates are higher among those who use nicotine replacement correctly.

Smokers who attempt to quit without any help, any support group, any gum, or any patch tend to have limited success. A group of leading cessation researchers, led by Sheldon Cohen, reported a decade ago on the quit-rates of more than 5,000 smokers who attempted to quit with-

Patched Up

2 hours	time it takes for nicotine in nicotine patch to plateau in the bloodstream
44%	increase of daily nicotine intake with 4-mg nicotine patch, vs. 2-mg nicotine replacement
10–20%	those using nicotine gum who will still be using it 9 months after quitting smoking
98–99%	those using nicotine gum who will eventually quit using it

out help from a "change agent" (e.g., therapist, nurse, doctor, or support group), and without any personalized assistance. Some of the smokers received self-quit manuals and other printed materials through the mail or through their place of employment. The report responded to published assertions that those who quit by themselves meet with more success than those who have help quitting.

Using data from ten long-term studies, the research group determined that self-quitting is not a panacea, and that those attempting to quit by themselves have no greater success than those attending formal programs. Those who smoke more than a pack a day are less successful at self-quitting over the long-term than are those who are considered light smokers. Those who smoke the most tend to have the most difficulty quitting in any setting, with or without assistance.

In the current climate of smoking cessation aids, the question has shifted from a discussion of self-quitting versus group treatment to the choice between using or not using the available medications. In addressing this question, groups of scientists have offered some valuable guidance.

Guidelines for Quitting

Many tobacco users who go through years of quitting and relapsing find specific approaches that work for them. Until recently, however, approaches to treatment of nicotine dependence were anything but standard. Every program, whether it was designed by the American Heart Association or a local hospital's nursing staff, had a somewhat different twist. Even though those differences in approach are likely to remain, two sets of guidelines issued in 1996 now help both interventionists and smoking consumers identify a successful approach to cessation.

These sets of guidelines were compiled by two separate blue-ribbon panels. The more comprehensive guideline was assembled by a panel of 19 specialists assigned to the task by the Agency for Health Care Policy and Research (AHCPR), a Public Health Service entity established in 1989 to conduct and support health service research. The AHCPR guideline (titled *Smoking Cessation*) was based on extensive searches of scientific and medical research. Hundreds of research reports were reviewed and synthesized. When scientific literature was incomplete or inconsistent, the panel and their consultants recommended what they believed to be sound practices.

Physician and public health specialist Michael Fiore, chairman of the panel that drafted the guidelines, called their publication "a critical event." Addressing a 1996 conference introducing the guidelines to the scientific and clinical communities, he elaborated: "This document, for the first time, provides clinicians, administrators, and smokers alike with a definitive, research-based answer to the question: what actions are necessary to improve the likelihood of successful smoking cessation for people already addicted to tobacco?" He deemed the dissemination of the information a "defining [moment] in reaching a goal that finally appears achievable: the elimination of tobacco addiction from our society."

Within a few months of the AHCPR recommendations, the American Psychiatric Association published a complementary guideline focusing on three target populations: (1) smoking patients being seen by

15 – 20	years it takes, after quitting smoking, to reduce the risk of cancer and likelihood of mortality; they never reach that of a never-smoker
1 – 3	years it takes, after quitting smoking, to reduce risk of recurrent myocardial infarction, sudden death, and stroke to levels approximating those of a never-smoker
1	days it takes after quitting smoking for carbon monoxide levels to return to those of a nonsmoker

psychiatrists for disorders unrelated to nicotine use; (2) smokers whose initial attempts at cessation failed and who need intensive treatment; and (3) smoking psychiatric patients confined to inpatient units or residential facilities where smoking is not allowed.

The two sets of guidelines combine to outline the best that science can offer the tobacco user who wants to quit. Together, they provide the most comprehensive descriptions to date of what works and what doesn't. This does not mean that other approaches might not work, but rather that they have not yet been demonstrated to work as well as the methods the guidelines list. Also, it does not mean that every smoker who uses these approaches will be successful. To the contrary, only a portion of those who attempt to quit using even the best combination of methods will succeed in any given quit attempt. The guidelines recognize that the process of learning to quit can span many years. Most youth who start smoking try to quit within a year. Most successful quit attempts come after repeated failures. Those who succeed generally are those who have learned from their earlier attempts.

A physically fit professor once described how he lost a considerable amount of weight. One day, during a time when he was seriously overweight, he went to a health club to sit in the sauna. As he basked in the heat, he noticed, with a glance, a man sitting across from him. He was immediately appalled by how fat the man was. Then he realized that he was seeing his own reflection in a mirror. That unguarded look at himself prompted him to start an exercise program. Within months, he was running several miles a day. He dropped to an ideal weight and stayed there. That unexpected vision forced him to confront a reality he had been denying.

Many cultures and nations have not yet had such a jolt. It is true that public health advocates have made considerable headway, and that the world of tobacco politics shifts daily, sometimes hourly. Even so, the world has not yet taken that unexpected glance in the international mirror that would motivate serious changes in smoking prevalence. What science knows about quitting smoking has not yet been translated into programs, policies, practices, and norms. We know how to help people stop smoking. We know what constitutes effective strategies and interventions. However, they are not yet available to the general population, at least not uniformly.

The panel that developed the AHCPR guidelines noted that some 70 percent of smokers say they would like to quit and have tried at least once. Also, they stated that a physician's advice can motivate smokers to stop using tobacco. But between those demonstrated facts and the reality of patient-physician contact is a sizable disconnection. Only about half of current smokers say that they have ever been asked about their smoking status or encouraged to quit. Even fewer have been offered advice on how to do so successfully.

The panel listed four solid suggestions that would increase clinicians' impact on smokers:

- Smoking cessation interventions must be institutionalized, through changes in health care delivery.
- Insurance companies must reimburse patients and clinicians for smoking cessation counseling and medications.

- Clinicians must offer motivational interventions to smokers who are not yet committed to quitting.
- The health care system's standards of care must reflect an obligation to intervene in a timely and appropriate way with patients who smoke.

The central theme of their report is that brief, effective help for smokers should be offered at each clinical visit. Every time a patient visits a doctor, dentist, or other care provider, stop-smoking help should be provided. It need not be an intensive program, although that might be more effective for some patients. The panel recommended that every medical student and other clinician in training be educated in smoking cessation. This would not only "transmit essential treatment skills," the panel noted, "but also inculcate the belief that cessation treatment meets the standard of good practice."

Moreover, the AHCPR panel recommended including questions about smoking cessation treatment in licensing and certification exams for all clinical disciplines. They also proposed that specialty societies adopt a uniform standard of competence in smoking cessation. They even recommended that clinicians who smoke should enter treatment programs so that they can stop smoking permanently themselves. This is important because clinicians serve as models for their patients.

How would this work in the typical visit to a doctor or other clinician? First, the clinician would determine whether you smoke. A report of your current and former tobacco use would be listed as a "vital sign," along with blood pressure and body temperature. (The concept that smoking should be assessed as a "vital sign" has been described by researcher Michael Fiore.)

If you smoke, you would be strongly advised to quit. The message should be clear, strong, and personalized. The clinician would offer to help you, would explain why it is important that you quit, and would describe how your smoking affects your medical condition and affects your family members. The message and encouragement would come not only from the clinician, but also from the clinic staff.

The clinician would then determine your willingness to attempt quitting, and would provide a motivational intervention to encourage you to quit. It might be as simple as asking you a question: "Are you ready to make a quit attempt at this time?" If you are ready, the clinician would offer help. If you want more intensive treatment than the clinician offers, he or she would refer you to a treatment program, and then would follow up with you. If you aren't motivated to quit at that time, the clinician or the staff could perform any of several motivating interventions, including offering you information about the relevance of quitting smoking, explaining the risks of smoking, and discussing the rewards of quitting. If you still aren't motivated to quit, the message might be repeated every time you visit the clinic.

Once you are willing to quit, the clinician would ask you to set a date for quitting, and would help you prepare for quitting. This might involve encouraging you to tell your family, friends, and others about your pending quit date. You would need to remove cigarettes from your environment, and stop smoking in places you routinely smoke, such as in your car. The clinician might talk to you about your previous attempts at quitting, and might work with you to anticipate anything that could challenge your coming quit attempt.

If your clinician follows the advice of the AHCPR panel, he or she would recommend that you use nicotine replacement therapy, unless you have a medical condition that might preclude its use, such as pregnancy or cardiovascular disease. Even in those cases, your physician would help you decide whether the relative benefits and risks of nicotine replacement would warrant your using it. And your clinician would make it very clear that you must not use tobacco while you use nicotine replacement.

Does this sound intrusive? If a patient is only seeking medical help because of a skin rash or an ear infection, why should she have to put up with all that commentary about her smoking? Why would her physician risk losing business?

Let's just say that clinicians have an interest in seeing their patients survive.

Intensive Treatment

As with all other human variables, the intensity of addiction can vary dramatically from person to person. Regardless of strength of character or willpower, a tobacco user's capacity to quit can be influenced by many other factors as well. Some smokers are able to quit smoking on their own, particularly if they use nicotine replacement carefully. Some smokers have better success in an intensive treatment program. This could be an outpatient program in which they have either group or individual counseling, or could involve their participation in an inpatient setting where their condition is monitored frequently and they receive a wealth of instruction and support. Intensive treatment is effective across most groups, regardless of sex, race, ethnicity, or health condition such as pregnancy.

The AHCPR panel reported a strong relationship between the intensity of counseling and success in quitting. More intense counseling resulted in a higher rate of cessation, overall. Intensity can be increased by lengthening each counseling session or by increasing the number of sessions and the number of weeks of treatment.

Researchers have identified factors predisposing a smoker to relapse, including dependence on nicotine, existence of a psychiatric problem, and low motivation to quit. These factors can be used to the smoker's advantage if they become the basis for matching the smoker to a suitable treatment. An example of such "treatment matching," as the practice is known, could be providing a depressive smoker with antidepressant treatment such as bupropion, if the drug is an appropriate prescription for that patient.

Another way to provide intensive treatment is to involve a variety of care providers, including perhaps a physician, a nurse, a dentist, a psychologist, or a pharmacist. Each might provide a unique perspective that could help a smoker in a different way.

Additionally, a smoker could participate in group or individual counseling, or both. The most effective content of counseling sessions would involve problem solving and skills training, to help the smoker

Matters of Life and Death

one-sixth	proportion of all deaths attributed to tobacco use in developed countries
35 years	age at which smokers begin to have a higher death rate than nonsmokers
one-half	proportion of all smokers who will eventually die from smoking
28 minutes	amount of life expectancy lost for each pack of cigarettes smoked
25 years	years of life expectancy a typical pack-a-day smoker loses

deal with temptations and high-risk situations that will arise after quitting. Support during treatment also helps boost cessation success. In addition, a smoker's chances of quitting can be enhanced through learning several self-administered aversive techniques, such as rapid smoking. Smokers in intensive treatment also are likely to be encouraged to use nicotine replacement therapy, since it consistently enhances cessation rates, independent of any other therapy that accompanies it.

How exactly should such programs be tailored? The panel recommended at least four to seven treatment sessions lasting at least 20 to 30 minutes. The sessions should be offered for at least two weeks, preferably more than eight weeks. Either individual or group counseling can be effective. It is important to provide follow-up assessments as well. The content should deal with motivation to quit and with relapse prevention. Every smoker, except those with special circumstances, should be offered nicotine replacement.

What the Health Care Industry Can Do

Of course, the onus for quitting smoking rests on the smoker. Even so, health care administrators, insurers, and purchasers can do much to make quitting smoking more feasible, and to make help more accessible. The AHCPR panel suggested a variety of approaches the health care industry could implement to encourage quitting:

- Expect clinicians to assess tobacco use and to assist with cessation as part of their usual responsibilities.
- Have every clinic assess patients' tobacco use.
- Provide an environment that supports systematic cessation treatment.
- Dedicate staff to providing effective stop-smoking treatment, and assess the treatment in performance evaluations.
- Display leadership, craft policies, and provide resources to foster cessation.
- Ensure that all health plans offer treatment for nicotine addiction. This would involve offering tobacco-use cessation counseling by health care providers, cessation classes, prescriptions for nicotine replacement, and other services. At present, as few as 11 percent of health insurance carriers provide coverage for nicotine addiction treatment.
- Reimburse fee-for-service clinicians for delivering effective stop-smoking treatment.
- Include such treatment among the duties expected of clinicians on salary.
- Comply with regulations mandating that all areas of a hospital be entirely smoke-free.
- Educate hospital staff about nicotine withdrawal and cessation techniques.

Tobacco use is spread across most demographic groups. Both men and women smoke, although there are differences in how they smoke and how it affects them. People of various racial groups smoke, although the preferences may differ. Nonetheless, the AHCPR panel determined that their general guidelines are applicable for virtually all groups. Tobacco use causes disease and death across all demographic groups; cessation can be successful with all groups. Both men and women, for instance, can benefit from the same basic cessation treatment. Women may confront different problems in quitting, including a greater possibility of depression and greater concern about weight gain. Even so, the panel identified no consistent evidence of differences between men and women in their success in smoking cessation treatment.

Some groups may have particular needs, although the evidence is weak that these groups would benefit from specially tailored programs. Disease and death related to tobacco use are more common among some U.S. minority groups, including African-Americans, Native Americans, Alaskan Natives, Asians, Pacific Islanders, and Hispanics. Even so, little research has studied interventions tailored only for particular groups, and there is no evidence that tailored programs lead to higher quit rates. In fact, smoking cessation programs developed for the population at large have been effective with minority groups.

Some tailoring makes logical sense, such as providing self-help materials in a language the smoker understands, or offering culturally appropriate models or examples. The panel noted that little work has been done to identify hindrances to successful quitting in groups with such problems as low education levels or lack of access to medical care. Clearly, such topics need to be studied.

Pregnant women constitute one group of smokers who should be strongly encouraged to quit using tobacco. Since it is common for pregnant smokers to play down or deny their smoking, the panel recommended offering intensive counseling treatment to pregnant smokers. Obviously, quitting before conception or early in a pregnancy is preferable to quitting later, but quitting at any point can be beneficial. Since it also is common for women who quit during pregnancy to relapse back to smoking after the child is born, the panel recommended educating women about the risks of smoking around an infant or child.

The question of whether a pregnant woman should use nicotine replacement has never been studied in a clinical trial, but pregnant women who cannot stop smoking without the help of nicotine replacement might benefit from using it, since it poses a reduced risk to health. Nonetheless, it would be preferable for a pregnant woman not to be using nicotine at all. In any case, pregnant women who participate in cessation counseling during pregnancy have higher quit rates than those who do not have such help. Even minimal counseling may be beneficial.

Hospitalized patients who smoke are another group who require consideration. The panel recommended that hospital staff ask all new patients about their tobacco use. Smoking status would be listed on the admission problem list and on the discharge diagnosis. The panel proposed that hospitals help all smokers quit during their hospitalization, including giving them nicotine replacement when appropriate. Since continued tobacco use could disrupt a patient's recovery, it is vital that hospitalized patients attempt to quit. Research evidence indicates that stop-smoking interventions can help hospitalized patients quit.

All hospitals accredited by the Joint Commission on Accreditation of Healthcare Organizations (JCAHO) are now required to be smoke-free. This provides tobacco-using patients an opportunity to quit smoking in a smoke-free environment. Hospitalized smokers could experience nicotine withdrawal symptoms if they are not allowed to smoke in the hospital. As with other groups of smokers, nicotine replacement may be appropriate for these patients.

Conditions such as depression, substance abuse, or other psychiatric problems are more common among smokers than among the general population. Between a third and half of patients seeking help to stop smoking may have a history of depression. At least one-fifth of those seeking help may have a history of problematic alcohol use. Going through nicotine withdrawal may worsen such conditions, and may put such individuals at greater risk for relapse. Nonetheless, smoking cessation treatment can help them quit and stay quit. Since the presence or absence of nicotine can affect how the body processes and uses some medications, such as antipsychotic drugs, clinicians need to monitor the effects and side effects of medications in smokers who are attempting to quit.

The experts reported that the best strategy is to quit smoking without attempting to diet at the same time. Once a smoker is confident that he or she will not relapse, and once the nicotine abstinence symptoms have passed, then it is appropriate to deal with the weight gain. Overall, what is most important is to maintain or establish a healthful lifestyle. The bottom line is this: Compared with the health risk of continuing smoking, any risk from postcessation weight gain is negligible.

The panel explained that dentists, in particular, are well positioned to help users of smokeless tobacco quit. Findings from a handful of research studies suggest that stop-smoking techniques are also effective with users of smokeless tobacco. It is possible that nicotine replacement may help smokeless tobacco users quit, although this has not been studied thoroughly.

Tobacco is not only an adult problem. Many grade-school children and adolescents appear to be as dependent on nicotine as adults are. Even so, in many communities little help exists for pediatric smokers. Cessation in young persons has not been studied extensively; consequently, the effectiveness of counseling and nicotine replacement for young people is undetermined. Some young people who attempt to quit using tobacco on their own quickly relapse. Clinicians such as physicians, nurses, and dentists can be an immediate source of help. The AHCPR panel suggested that clinicians consider nicotine replacement for young users of tobacco, as long as the degree of dependence and body weight are taken into consideration when determining dosage.

The panel's many recommendations were based on evidence of what works. As they emphasized, "an absence of studies should not be confused with a lack of efficacy." They did not report, for example, on the use of bupropion, since they conducted their analyses before bupropion was approved for smoking cessation treatment. In cases where evidence was thin or inconclusive, the panel rendered no opinion about a treatment approach. Sometimes, data were inadequate or unavailable, as in a comparison of nicotine patch and nicotine gum. No adequate, published studies had compared the two methods; consequently, the panel did not compare them. The panel also avoided ranking therapies in order of superiority. They did note, however, that longer person-to-person therapeutic interactions have a greater impact than minimal contact, and are superior to having no contact.

A Process

Stephen King's tale *Cat's Eye* portrayed a stop-smoking method that could make even the most addicted smoker throw away the cigarettes for good. A smoker who agreed to participate in the stop-smoking program and then sneaked a cigarette had much to lose, including a finger or a family member. Effective as draconian measures may be, they are neither legal nor necessary. Stopping smoking may take repeated tries, but it is feasible. The secret is in learning to quit.

Quitting smoking is a process, not a single event. Many users of tobacco find that they must learn to quit before they can succeed permanently. Learning to quit can involve learning to manage lapse and relapse episodes, turning them from catastrophe to beneficial experience. Some tobacco users are more successful in quitting when they focus on reducing their risk for relapse. Ex-smokers learn that they are tempted to relapse in certain situations. The "triggers" that lead them back to smoking are individual. For some, just going to a gathering where others will be smoking is enough to trigger a relapse. For others, it is the quiet moment after a meal, or an unexpected high-stress crisis at work.

Hundreds of scientific studies report that a host of factors influence success in quitting, just as many individual factors lead to tobacco use. Although first-try failures are discouraging and common, the odds are high that a smoker determined to quit can eventually do so. Many have, and many more can.

References

The text and tables in this book are drawn from hundreds of journal articles, chapters, books, newspaper and magazine reports, and other sources. Those listed below are cited specifically in the text, or are key articles useful for understanding an author's cited works.

Adler, L. E., L. Hoffer, J. M. Griffith, M. C. Waldo, and R. Freedman. 1992. Normalization by nicotine of deficient auditory sensory gating in the relatives of schizophrenics. Biological Psychiatry 32:607-616.

Adler, L. E., L. Hoffer, A. Wise, and R. Freedman. 1993. Normalization of auditory physiology by cigarette smoking in schizophrenic patients. American Journal of Psychiatry 150:1856-1861.

Alexander, J., and P. Alexander. 1994. Gender differences in tobacco use and the commodification of tobacco in Central Borneo. Social Science and Medicine 38: 603-608.

American Psychiatric Association. 1987. Diagnostic and Statistical Manual of Mental Disorders (3rd ed., rev.). Washington, DC.

Arcavi L., P. Jacob 3rd, M. Hellerstein, and N. L. Benowitz. 1994. Divergent tolerance to metabolic and cardiovascular effects of nicotine in smokers with low and high levels of cigarette consumption. Clinical Pharmacology & Therapeutics 56:55-64.

Arendash, G. W., P. R. Sanberg, and G. J. Sengstock. 1995. Nicotine enhances the learning and memory of aged rats. Pharmacology Biochemistry and Behavior 52:517-523.

Audrain, J., R. Klesges, and L. Klesges. 1995. Relationship between obesity and the metabolic effects of smoking in women. Health Psychology 14:116-123.

Bahuchet, S., ed. 1994. [The Situation of Indigenous Peoples in Tropical Forests]. Brussels: Commission des Communautés Européenes.

Balbach, E. D., and S. A. Glantz. 1995. Tobacco information in two grade school newsweeklies: A content analysis. American Journal of Public Health 85:1650-1653.

Balfour, D. J. K. 1994. Neural mechanisms underlying nicotine dependence. Addiction 89:1419-1423.

Barendregt, J. J., L. Bonneaux, and P. J. van der Maas. 1997. The health care costs of smoking. The New England Journal of Medicine 337:1052-1057.

Barnum, H. 1994. The economic burden of the global trade in tobacco. Tobacco Control 3:358-361.

Bauman, K. E., R. L. Flewelling, and J. LaPrelle. 1991. Parental cigarette smoking and cognitive performance of children. Health Psychology 10:282-288.

Benet, S. V. 1928. John Brown's Body. Garden City, NY: Doubleday, Doran and Co.

Bernstein, M., A. Morabia, S. Heritier, and N. Katchatrian. 1996. Passive smoking, active smoking, and education: Their relationship to weight history in women in Geneva. American Journal of Public Health 86:1267-1272.

Bickel, W. K., and R. J. DeGrandpre (eds). 1996. Drug Policy and Human Nature: Psychological Perspectives on the Prevention, Management, and Treatment of Illicit Drug Abuse. New York: Plenum.

Blaze-Temple, D., and S. K. Lo. 1992. Stages of drug use: A community survey of Perth teenagers. British Journal of Addiction 87:215-225.

Bolinder, G. 1997. Smokeless tobacco—a less harmful alternative? In The Tobacco Epidemic, Fagerström, K. O., ed. Basel, Switzerland: Karger.

Botvin, G. J., E. Baker, C. J. Goldberg, L. Dusenbury, and E. M. Botvin. 1992. Correlates and predictors of smoking among black adolescents. Addictive Behaviors 117:97-103.

Botvin, G. J., J. A. Epstein, S. Schinke, and T. Diaz. 1994. Predictors of cigarette smoking among inner-city minority youth. Experimental and Behavioral Pediatrics 15:67-73.

Bradbury, R. 1985. Dandelion Wine. Reissue ed. New York: Bantam-Spectra.

Breslau, N., M. M. Kilbey, and P. Andreski. 1993. Nicotine dependence and major depression: New evidence from a prospective investigation. Archives of General Psychiatry 50:31-35.

Breslau, N. 1995. Psychiatric comorbidity of smoking and nicotine dependence. Behavior Genetics 25:95-101.

Breslau, N., and E. L. Peterson. 1996. Smoking cessation in young adults: age at initiation of cigarette smoking and other suspected influences. American Journal of Public Health, 86: 214-220.

Breslau, N., E. L. Peterson, L. R. Schultz, H. D. Chilcoat, P. Andreski. 1998. Major depression and stages of smoking. Archives of General Psychiatry 55:161-166.

Brigham, J., J. Gross, M. L. Stitzer, and L. J. Felch. 1994. Effects of a restricted work-site smoking policy on employees who smoke. American Journal of Public Health 84:773-778.

Burns, D. M. 1992. Positive evidence on effectiveness of selected smoking prevention programs in the United States. Monographs—National Cancer Institute 12:17-20.

California Center for Health Improvement. 1997. Environmental tobacco smoke and lung cancer risk. www.webcom.com/cchi/PUB/tobaccosmoke.html

California Environmental Protection Agency. 1997. Health Effects of Exposure to Environmental Tobacco Smoke: Final Draft for Scientific, Public, and SRP Review. www.calepa.cahwnet.gov/oehha/docs/ets/ets-main.htm

Camp, D. E., R. C. Klesges, and G. Relyea. 1993. The relationship between body weight concerns and adolescent smoking. Health Psychology 12:24-32.

Cardador, M. T., A. R. Hazan, and S. A. Glantz. 1995. Tobacco industry smokers' rights publications: A content analysis. American Journal of Public Health 85:1212-1217.

Carmelli, D., G. E. Swan, and D. Robinette. 1993. The relationship between quitting smoking and changes in drinking in World War II veteran twins. Journal of Substance Abuse 5:103-116.

Caskey, N., W. C. Wirshing, M. E. Jarvik, D. C. Madsen, and J. L. Elms. 1997. Smoking influences on symptoms in schizophrenics. Paper presented at the third Annual Meeting of the Society for Research on Nicotine and Tobacco, Nashville, TN, June.

Centers for Disease Control. See Massachusetts Medical Society.

Charlton, A. 1995. Children and tobacco. Tobacco Control 4:103-104.

Chassin, L., C. C. Presson, J. S. Rose, and S. J. Sherman. 1996. The natural history of cigarette smoking from adolescence to adulthood: Demographic predictors of continuity and change. Health Psychology 15:478-484.

Chen, K., and D. B. Kandel. 1995. The natural history of drug use from adolescence to the mid-thirties in a general population sample. American Journal of Public Health 85:41-47.

Cohen, C., W. B. Pickworth, and J. E. Henningfield. 1991. Cigarette smoking and addiction. Clinics in Chest Medicine 12:701-710.

Cohen, S., E. Lichtenstein, J. O. Prochaska, J. S. Rossi, E. R. Gritz, C. R. Carr, C. T. Orleans, V. J. Schoenbach, L. Biener, D. Abrams, C. DiClemente, S. Curry, G. A. Marlatt, K. M. Cummings, S. L. Emont, G. Giovino, and D. Ossip-Klein. 1989. Debunking myths about self-quitting: Evidence from 10 prospective studies of persons who attempt to quit smoking by themselves. American Psychologist 44:1355-1365.

Cronk, C.E., and P. D. Sarvela. 1997. Alcohol, tobacco, and other drug use among rural/small town and urban youth: a secondary analysis of the monitoring the future data set. American Journal of Public Health 87:760-764.

Davis, R. M. 1992. The language of nicotine addiction: Purging the word "habit" from our lexicon. Tobacco Control 1:163-164.

Eisner, J. R., M. Morgan, and P. Gammage. 1987. Belief correlates of perceived addiction in young smokers. European Journal of Psychology of Education 2:307-310.

Eliasson, B., M. R. Taskinen, and U. Smith. 1996. Long-term use of nicotine gum is associated with hyperinsulinemia and insulin resistance. Circulation 94:878-881.

Elrod, K., J. J. Buccafusco, and W. J. Jackson. 1988. Nicotine enhances delayed matching-to-sample performance by primates. Life Sciences 43:277-287.

Fiore, M. C. 1991. The new vital sign: Assessing and documenting smoking status. The Journal of the American Medical Association 266:3183-3184.

Fiore, M. C. 1997. AHCPR smoking cessation guideline: A fundamental review. Tobacco Control 6(Suppl. 1):S4-S8.

Fischer, P. M., M. P. Schwartz, J. W. Richards Jr., A. O. Goldstein, T. Rojas. 1991. Brand logo recognition by children aged 3 to 6 years: Mickey Mouse and Old Joe the Camel. The Journal of the American Medical Association 266:3145-3148.

Flegal, K. M., R. P. Troiano, E. R. Pamuk, R. J. Kuczmarski, and S. M. Campbell. 1995. The influence of smoking cessation on the prevalence of overweight in the United States. New England Journal of Medicine 333:1165-1170.

Freedman, R., H. Coon, M. Myles-Worsley, A. Orr-Urtreger, A. Olincy, A. Davis, M. Polymeropoulos, J. Holik, J. Hopkins, M. Hoff, J. Rosenthal, M. C. Waldo, F. Reimherr, P. Wender, J. Yaw, D. A. Young, C. R. Breese, C. Adams, D. Patterson, L. E. Adler, L.

Kruglyak, S. Leonard, and W. Byerley. 1997. Linkage of a neuro-physiological deficit in schizophrenia to a chromosome 15 locus. Proceedings of the National Academy of Sciences 94:587-592.

Frost, K., E. Frank, and E. Maibach. 1997. Relative risk in the news media: A quantification of misrepresentation. American Journal of Public Health 87:842-845.

Gammon, M. D., J. Schoenberg, H. Ahsan, H. Risch, T. Vaughan, W. Chow, H. Rotterdam, A. West, R. Dubrow, J. Stanford, S. Mayne, D. Farrow, S. Niwa, W. Blot, and J. J. Fraumeni. 1997. Tobacco, alcohol, and socioeconomic status and adenocarcinomas of the esophagus and gastric cardia. Journal of the National Cancer Institute 89:1277-1284.

Gilbert, D. G. 1995. The Situation x Trait Adaptive Response (STAR) model of nicotine's effects on physiological, emotional, and information processing states. Psychophysiology 32:S7.

Gilbert, D. G. 1995. Smoking: Individual Differences, Psychopathology, and Emotion. Washington, D.C.: Taylor and Francis.

Glover, E. D., M. Laflin, D. Flannery, and D. L. Albritton. 1989. Smokeless tobacco use among American college students. Journal of American College Health 38:81-85.

Gold, M. S. 1995. Tobacco. New York: Plenum Medical Book Company.

Gori, G. B. 1995. Policy against science: The case of environmental tobacco smoke. Risk Analysis 15:15-22.

Greene, J. M., S. T. Ennett, and C. L. Ringwalt. 1997. Substance use among runaway and homeless youth in three national samples. American Journal of Public Health 87:229-235.

Gross, T. M., M. E. Jarvik, and M. R. Rosenblatt. 1993. Nicotine abstinence produces content-specific Stroop interference. Psychopharmacology 110:333-336.

Groth-Marnat, G., S. Leslie, and M. Renneker. 1996. Tobacco control in a traditional Fijian village: Indigenous methods of smoking cessation and relapse prevention. Social Science and Medicine 43:473-477.

Grunberg, N. E. 1991. Smoking cessation and weight gain. The New England Journal of Medicine 324:768-769.

Haaga, D. A., M. M. Gillis, and W. McDermut. 1993. Lay beliefs about the causes and consequences of smoking cessation maintenance. The International Journal of the Addictions 28:369-375.

Hall, S. M., D. Ginsberg, and R. T. Jones. 1986. Smoking cessation and weight gain. Journal of Consulting and Clinical Psychology 54:342-346.

Hall, S. M., R. F. Muñoz, V. I. Reus, and K. L. Sees. 1993. Nicotine, negative affect, and depression. Journal of Consulting and Clinical Psychology 61:761-767.

Hall, S. M., C. D. Tunstall, K. L. Vila, and J. Duffy. 1992. Weight gain prevention and smoking cessation: Cautionary findings. American Journal of Public Health 82:799-803.

Hatsukami, D. K., D. Anton, A. Callies, and R. Keenan. 1991. Situational factors and patterns associated with smokeless tobacco use. Journal of Behavioral Medicine 14:383-396.

Hazan, A. R., and S. A. Glantz. 1995. Current trends in tobacco use on prime-time fictional television. American Journal of Public Health 85:116-117.

Hazan, A. R., H. L. Lipton, and S. A. Glantz. 1994. Popular films do not reflect current tobacco use. American Journal of Public Health 84:998-1000.

Heishman, S. J., F. R. Snyder, and J. E. Henningfield. 1993. Performance, subjective, and physiological effects of nicotine in non-smokers. Drug and Alcohol Dependence 34:11-18.

Heishman, S. J. 1996. Effects of nicotine on cognitive and psychomotor abilities in nonsmokers. Paper presented at Smoking, Nicotine, and Human Performance, Washington, D.C., November.

Henningfield, J. E. 1996. Public policy: up in smoke—nicotine research travails. Paper presented at the annual meeting of the American Psychological Association, Toronto, Canada, August.

Henningfield, J. E., M. Hariharan, and L. T. Kozlowski. 1996. Nicotine content and health risks of cigars. The Journal of the American Medical Association 276:1857-1858.

Henningfield, J. E., J. Slade, L. Biener, S. Leischow, J. Henningfield, and M. Stitzer. 1997. Reducing tobacco-caused death and disease by reducing exposure in cigarette smokers. Symposium presented at the Fourth Annual Meeting of the Society for Research on Nicotine and Tobacco, Nashville, Tennessee, June.

Henningfield, J. E., R. V. Fant, A. Radzius, S. Frost, and W. Rickert. 1998. Nicotine delivery and smoke pH of seventeen cigar brands. Paper presented at the Fifth Annual Meeting of the Society for Research on Nicotine and Tobacco, New Orleans, Louisiana, March.

Howard, G., L. E. Wagenknecht, G. L. Burke, A. Diez-Roux, G. W. Evans, P. McGovern, F. J. Nieto, and G. S. Tell. 1998. Cigarette smoking and progression of atherosclerosis: The Atherosclerosis Risk in Communities (ARIC) Study. The Journal of the American Medical Association 279:119-124.

Huber, G. L., and R. J. Pandina. 1997. The economics of tobacco use. In The Tobacco Epidemic, K. O. Fagerström, ed. Basel, Switzerland: Karger.

Hughes, J. R., S. T. Higgins, and W. K. Bickel. 1994. Nicotine withdrawal versus other drug withdrawal syndromes: Similarities and dissimilarities. Addiction 89:1461-1470.

Hurt, R. D., R. E. Finlayson, R. M. Morse, and L. J. Davis. 1988. Alcoholism in elderly persons: Medical aspects and prognosis of 216 inpatients. Mayo Clinic Proceedings 63:753-760.

Hurt, R. D., D. P. Sachs, E. D. Glover, K. P. Offord, J. A. Johnston, L. C. Dale, M. A. Khayrallah, D. R. Schroeder, P. N. Glover, C. R. Sullivan, I. T. Croghan, and P. M. Sullivan. 1997. A comparison of sustained-release bupropion and placebo for smoking cessation. The New England Journal of Medicine 337:1195-1202.

Jackson, C., L. Henriksen, D. Dickinson, and D. W. Levine. 1997. The early use of alcohol and tobacco: Its relation to children's competence and parents' behavior. American Journal of Public Health 87:359-364.

Jinot, J., S. Bayard. 1995. Environmental tobacco smoke: Science vs. rhetoric. Risk Analysis 15:91-96.

Jones, G. M. M., B. J. Sahakian, R. Levy, D. M. Warburton, and J. A. Gray. 1992. Effects of acute subcutaneous nicotine on attention, information processing and short-term memory in Alzheimer's disease. Psychopharmacology 108:485-494.

Jones, T. 1963. The Fantasticks. Milwaukee, Wisconsin: Hal Leonard.

Kawachi, I., G. A. Colditz, F. E. Speizer, J. E. Manson, M. J. Stampfer, W. C. Willett, and C. H. Hennekens. 1997. A prospective study of passive smoking and coronary heart disease. Circulation 95:2374-2379.

Kendler, K. S., M. C. Neale, C. J. MacLean, A. C. Heath, L. H. Eaves, and R. C. Kessler. 1993. Smoking and major depression: A causal analysis. Archives of General Psychiatry 50:36-43.

Kessler, D. A. 1994. Statement on nicotine-containing cigarettes. Tobacco Control 3:148-158.

Kirch, D.G., Bigelow, L.B., Weinberger, D.R., Lawson, W.B., Wyatt, R.J. 1985. Polydipsia and chronic hyponatremia in schizophrenic inpatients. Journal of Clinical Psychiatry, 46:179-181.

Klein, M. 1997. Holes in the smoke screen. American Demographics. July, p. 31.

Klein, R. 1993. Cigarettes Are Sublime. Durham, NC: Duke University Press.

Klesges, R. C., and L. M. Klesges. 1988. Cigarette smoking as a dieting strategy in a university population. International Journal of Eating Disorders 7:413-419.

Klesges, R. C., S. E. Winders, A. W. Meyers, L. H. Eck, K. D. Ward, C. M. Hultquist, J. W. Ray, and W. R. Shadish. 1997. How much weight gain occurs following smoking cessation? A comparison of weight gain using both continuous and point prevalence abstinence. Journal of Consulting and Clinical Psychology 65:286-291.

Koop, C. E. 1997. The truth about secondhand smoke. The Washington Post. 11 October, p. A25.

Kozlowski, L. T., J. L. Pillitteri, B. A. Yost, M. E. Goldberg, and F. M. Ahern. 1998. Advertising fails to inform smokers of official tar and nicotine yields of cigarettes. Journal of Applied Biobehavioral Research 3:55-64.

Le Houezec, J., R. Halliday, N. L. Benowitz, E. Callaway, H. Naylor, and K. Herzig. 1994. A low dose of subcutaneous nicotine improves information processing in non-smokers. Psychopharmacology 114:628-634.

Leistikow, B., D. Martin, and J. Jacobs. 1997. Smoking as a risk factor for suicide: A meta-analysis. Paper presented at the Third Annual Meeting of the Society for Research on Nicotine and Tobacco, Nashville, Tennessee, June.

Leonard, S., C. Adams, C. R. Breese, L. E. Adler, P. Bickford, W. Byerley, H. Coon, J. M. Griffith, C. Miller, M. Myles-Worsley, H. T. Nagamoto, Y. Rollins, K. E. Stevens, M. Waldo, and R. Freedman. 1996. Nicotinic receptor function in schizophrenia. Schizophrenia Bulletin 22:431-445.

Levin, E. D., S. J. Briggs, N. C. Christopher, and J. E. Rose. 1993. Prenatal nicotine exposure and cognitive performance in rats. Neurotoxicology and Teratology 15:251-260.

Maag, J. W., D. M. Irvin, R. Reid, and S. F. Vasa. 1994. Prevalence and predictors of substance use: A comparison between adolescents with and without learning disabilities. Journal of Learning Disabilities 27:223-234.

Marlatt, G. A. 1985. Relapse prevention: Theoretical rationale and review of the model. In Relapse Prevention: Maintenance Strategies in the Treatment of Addictive Behaviors, G. A. Marlatt and J. R. Gordon, eds. New York: Guilford.

Massachusetts Medical Society. 1996. Projected smoking-related deaths among youth—United States. Morbidity and Mortality Weekly Report 45:971-974.

Massachusetts Medical Society. 1996. Tobacco use and usual source of cigarettes among high school students—United States, 1995. Morbidity and Mortality Weekly Report 45:413-418.

Massachusetts Medical Society. 1997. Cigar smoking among teenagers—United States, Massachusetts, and New York, 1996. Morbidity and Mortality Weekly Report 46:433-440.

Massachusetts Medical Society. 1997. State- and sex-specific prevalence of selected characteristics—Behavioral Risk Factor Surveillance System, 1994 and 1995. Morbidity and Mortality Weekly Report 46:1-10.

Massachusetts Medical Society. 1998. Tobacco use among high school Students–United States, 1997. Morbidity and Mortality Weekly Report, 3 April 1998.

Mâsse, L. C., and R. E. Tremblay. 1997. Behavior of boys in kindergarten and the onset of substance use during adolescence. Archives of General Psychiatry 54:62-68.

McKenna, F. P., D. M. Warburton, and M. Winwood. 1993. Exploring the limits of optimism: The case of smokers' decision making. British Journal of Psychology 84:389-394.

Meier, K. J., and M. J. Licari. 1997. The effect of cigarette taxes on cigarette consumption, 1955 through 1994. American Journal of Public Health 87:1126-1130.

Meyers, A. W., R. C. Klesges, S. E. Winders, K. D. Ward, B. A. Peterson, and L. H. Eck. 1997. Are weight concerns predictive of smoking cessation? A prospective analysis. Journal of Consulting and Clinical Psychology 65:448-452.

Miller, N. S. 1991. Nicotine addiction as a disease. In The Clinical Management of Nicotine Dependence, J. A. Cocores, ed. New York: Springer-Verlag.

Murray, B. 1997. Why aren't antidrug programs working? American Psychological Association Monitor. September, p. 30.

National Cancer Institute. 1998. Cigars: Health Effects and Trends, David M. Burns (ed.), Bethesda, Maryland: National Cancer Institute.

National Health and Medical Research Council. 1997. The health effects of passive smoking: A scientific information paper. Canberra, Australia: Australian Government Publishing Service. www.health.gov.au/hfs/nhmrc/advice/nhmrc/foreword.htm

Nelson, D. R. 1997. Pretension aside, a cigar is just a cigar. Puget Sound Business Journal. 1 September.

Newhouse, P. A., A. Potter, and E. D. Levin. 1997. Nicotinic system involvement in Alzheimer's and Parkinson's diseases: Implications for therapeutics. Drugs and Aging 11:206-228.

Nothwehr, F., H. Lando, J. K. Bobo. 1995. Alcohol and tobacco use in the Minnesota Heart Health Program. Addictive Behaviors 20:463-470.

Pennington, J., and J. Tate. 1998. Why is unrealistic optimism extended to close others who smoke? Paper presented at the Fourth Annual Meeting of the Society for Research on Nicotine and Tobacco, New Orleans, Louisiana, March.

Perkins, K. A. 1996. Sex differences in nicotine versus nonnicotine reinforcement as determinants of tobacco smoking. Experimental and Clinical Psychopharmacology 4:166-177.

Perkins, K. A. 1997. Nicotine effects during physical activity. In Esteve Foundation Symposium VII: The Clinical Pharmacology of Sport and Exercise, T. Reilly and M. Orme, eds., Amsterdam: Elsevier.

Perkins, K. A., A. DiMarco, J. E. Grobe, A. Scierka, R. L. Stiller. 1994. Nicotine discrimination in male and female smokers. Psychopharmacology 116:407-413.

Perkins, K. A., J. Rohay, E. N. Meilahn, R. R. Wing, K. A. Mathews, L. H. Kuller. 1993. Diet, alcohol, and physical activity as a function of smoking status in middle-aged women. Health Psychology 12:410-415.

Perkins, K. A., M. Levine, M. D. Marcus, and S. Shiffman. 1997. Addressing women's concerns about weight gain due to smoking cessation. Journal of Substance Abuse Treatment 14:1-10.

Perkins, K. A., J. E. Sexton, L. H. Epstein, A. DiMarco, C. Fonte, R. L. Stiller, A. Scierka, and R. G. Jacob. 1994. Acute thermogenic effects of nicotine combined with caffeine during light physical activity in male and female smokers. American Journal of Clinical Nutrition 60:312-319.

Perkins, K. A., J. E. Sexton, and A. DiMarco. 1996. Acute thermogenic effects of nicotine and alcohol in healthy male and female smokers. Physiology and Behavior 60:305-309.

Pierce, J. P., W. S. Choi, E. A. Gilpin, A. J. Farkas, and C. C. Berry. 1998. Tobacco industry promotion of cigarettes and adolescent smoking. The Journal of the American Medical Association 279: 511-515.

Pierce, J. P., L. Lee, E. A. Gilpin. 1994. Smoking initiation by adolescent girls, 1944 through 1988: An association with targeted advertising. The Journal of the American Medical Association 271:608-611.

Pierce, J. P., and E. Gilpin. 1996. How long will today's new adolescent smoker be addicted to cigarettes? American Journal of Public Health 86:253-256.

Pomerleau, C. S., and O. F. Pomerleau. 1992. Euphoriant effects of nicotine in smokers. Psychopharmacology 108:460-465.

Pomerleau, C. S., E. Ehrlich, J. C. Tate, J. L. Marks, K. A. Flessland, and O. F. Pomerleau. 1993. The female weight-control smoker: A profile. Journal of Substance Abuse 5:391-400.

Pomerleau, O. F. 1992. Nicotine and the central nervous system: Biobehavioral effects of cigarette smoking. The American Journal of Medicine 93:1A-2S - 1A-7S.

Prochaska, J. O., and C. C. DiClemente. 1992. Stages of change in the modification of problem behaviors. Progress in Behavior Modification 28:183-218.

Rachlin, H. 1995. Behavioral economics without anomalies. Journal of the Experimental Analysis of Behavior 64:397-404.

Raeburn, P., and G. DeGeorge. 1997. You bet I mind if you smoke. Business Week, 15 September 1997.

Rásky É.; W. J. Stronegger, W. Freidl. 1996. The relationship between body weight and patterns of smoking in women and men. International Journal of Epidemiology 25: 1208-1212.

Redhead, C. S., and R. E. Rowberg. 1995. Environmental Tobacco Smoke and Lung Cancer Risk. Washington, D.C.: Congressional Research Service.

Reid, D. J., A. D. McNeill, and T. J. Glynn. 1995. Reducing the prevalence of smoking in youth in Western countries: An international review. Tobacco Control 4:266-277.

Resnick, M. P. 1993. Treating nicotine addiction in patients with psychiatric co-morbidity. In Nicotine Addiction: Principles and Management, C. T. Orleans and J. Slade, eds. New York: Oxford University Press.

Restak, R. M. 1996. Brainscapes: An Introduction to What Neuroscience Has Learned About the Structure, Function, and Abilities of the Brain. New York: Hyperion.

Rigotti, N. A., J. R. DiFranza, Y. Chang, T. Tisdale, B. Kemp, and D. E. Singer. 1997. The effect of enforcing tobacco-sales laws on adolescents' access to tobacco and smoking behavior. New England Journal of Medicine 337:1044-1051.

Robinson, J. H., and W. S. Pritchard. 1992. The meaning of addiction: Reply to West. Psychopharmacology 108:411-416.

Robinson, J. H., and W. S. Pritchard. 1992. The role of nicotine in tobacco use. Psychopharmacology 108:397-407.

Rohde, P., P. M. Lewinsohn, and J. R. Seeley. 1994. Are adolescents changed by an episode of major depression? Journal of the American Academy of Child and Adolescent Psychiatry 33:1289-1298.

Rotfeld, H. J. 1996. Tobacco firms are efficient marketers. Should they be? Marketing News. 9 September, pp.4, 11.

Royal Society of Canada. 1989. Tobacco, nicotine, and addiction. Ottawa: T & H Printers Ltd.

Russell, M. A. H. 1990. Nicotine intake and its control over smoking. In Nicotine Psychopharmacology: Molecular, Cellular, and

Behavioural Aspects, S. Wonnacott, M. A. H. Russell, and I. P. Stolerman, eds. Oxford: Oxford University Press.

Russell, M. A. H. 1990. The nicotine addiction trap: A 40-year sentence for four cigarettes. British Journal of Addiction 85:293-300.

Samuelson, R. J. 1997. Do Smokers Have Rights? Washington Post, 24 September 1997, A21.

Schlapman, N. 1987. Developing a workplace smoking policy. AAOHN Journal 35:337-339.

Scruton, R. 1998. Anything goes—except smoking. Wall Street Journal, 9 February 1998, A18.

Sherwood, N., J. S. Kerr, and I. Hindmarch. 1992. Psychomotor performance in smokers following single and repeated doses of nicotine gum. Psychopharmacology 108:432-436.

Sherwood, N. 1995. Effects of Cigarette Smoking on Performance in a Simulated Driving Task. Neuropsychobiology 32.

Shiffman, S. 1989. Tobacco "chippers"–individual differences in tobacco dependence. Psychopharmacology 97:539-547.

Shumaker, S. A., and N. E. Grunberg. 1986. Proceedings of the national working conference on smoking relapse. Health Psychology 5 (Suppl.).

Slade, J. 1997. Historical notes on tobacco. In The Tobacco Epidemic, K. O. Fagerström, ed. Basel, Switzerland: Karger.

Snyder, F. R., F. C. Davis, and J. E. Henningfield. 1989. The tobacco withdrawal syndrome: Performance decrements assessed on a computerized test battery. Drug and Alcohol Dependence 23:259-266.

Spilich, G. J. 1994. Cognitive benefits of nicotine: Fact or fiction? Addiction 89:141-142.

Stockwell, T. F., and S. A. Glantz. 1998. Tobacco use is increasing in popular film. Tobacco Control 7:282-284.

Sussman, S., C. W. Dent, L. A. McAdams, A. W. Stacy, D. Burton, and B. R. Flay. 1994. Group self-identification and adolescent cigarette smoking: A 1-year prospective study. Journal of Abnormal Psychology 103:576-580.

Swan, G. E., and D. Carmelli. 1995. Characteristics associated with excessive weight gain after smoking cessation in men. American Journal of Public Health 86:73-77.

Swan, G. E., D. Carmelli, and L. R. Cardon. 1996. The consumption of tobacco, alcohol, and coffee in caucasian male twins: A multivariate genetic analysis. Journal of Substance Abuse 8:19-31.

Swan, G. E., M. M. Ward, and L. M. Jack. 1996. Abstinence effects as predictors of 28-day relapse in smokers. Addictive Behaviors 21:481-490.

Swan, G. E., D. Carmelli, and L. R. Cardon. 1997. Heavy consumption of cigarettes, alcohol, and coffee in male twins. Journal of Studies on Alcohol 58:182-190.

Sweanor, D. T. 1997. Regulation of tobacco and nicotine. In The Tobacco Epidemic, K. O. Fagerström, ed. Basel, Switzerland: Karger.

Tillgren, P., B. J. Haglund, H. Gilljam, and L. E. Holm. 1992. A tobacco quit and win model in the Stockholm cancer prevention programme. European Journal of Cancer Prevention 1:361-366.

U. S. Department of Health and Human Services. 1988. The Health Consequences of Smoking: Nicotine Addiction. Washington, D.C.: U.S. Government Printing Office.

U. S. Department of Health and Human Services. 1990. The health benefits of smoking cessation: A report of the surgeon general. Washington, D.C.: U.S. Government Printing Office.

U. S. Department of Health and Human Services. 1996. Smoking Cessation. Clinical Practice Guideline No. 18. Washington, D.C.: Agency for Health Care Policy and Research.

U. S. Department of Health, Education, and Welfare. 1964. Smoking and Health: Report of the Advisory Committee to the Surgeon General of the public Health Service. DHEW Publication No.(PHS) 1103.

U. S. Environmental Protection Agency. 1992. Respiratory Health Effects of Passive Smoking: Lung Cancer and Other Disorders. Washington, D.C.: Office of Health and Environmental Assessment, Office of Research and Development.

Viscusi, W. K. 1992. Smoking: Making the Risky Decision. New York: Oxford University Press.

Warburton, D. M., A. D. Revell, and D. H. Thompson. 1991. Smokers of the future. British Journal of Addiction 86:621-625.

Warburton, D. M., K. Wesnes, K. Shergold, and M. James. 1986. Facilitation of learning and state dependency with nicotine. Psychopharmacology 89:55-59.

Warner, K. E., L. M. Goldenhar, and C. G. McLaughlin. 1992. Cigarette advertising and magazine coverage of the hazards of smoking: A statistical analysis. New England Journal of Medicine 326:305-309.

Weekley, C. K., R. C. Klesges, and G. Relyea. 1992. Smoking as a weight-control strategy and its relationship to smoking status. Addictive Behaviors 17:259-271.

West, R., and M. Gossop. 1994. Overview: A comparison of withdrawal symptoms from different drug classes. Addiction 89.

Wiebel, F. J. 1997. The health effects of passive smoking. In The Tobacco Epidemic, Fagerström, K. O., ed. Basel, Switzerland: Karger.

Williamson, D. F., J. Madans, R. F. Anda, J. C. Kleinman, G. A. Giovino, and T. Byers. 1991. Smoking cessation and severity of weight gain in a national cohort. New England Journal of Medicine 324:739-745.

Wilson, A. L., L. K. Langley, J. Monley, T. Bauer, S. Rottunda, E. McFalls, C. Kovera, and J. R. McCarten. 1995. Nicotine patches in Alzheimer's disease: Pilot study on learning, memory, and safety. Pharmacology Biochemistry and Behavior 51:509-514.

Winter, J., ed. (In press.) Deer Person's Gift: Tobacco Use by Native North Americans.

Wong, P. P., and A. Bauman. 1997. How well does epidemiological evidence hold for the relationship between smoking and adverse obstetric outcomes in New South Wales? Australian & New Zealand Journal of Obstetrics & Gynaecology 37:168-173.

World Health Organization. 1995. Tobacco: a global emergency. Tobacco Control 4:297-298.

Index

Adler, Lawrence E., 119
adolescents. *See* youth, tobacco use
adrenaline, as abstinence effect, 214
advertising, tobacco
  expenditures, 45
  female response to, 142
  foreign, 205
  limitations on, 8, 9, 19
  in magazines, 22-23
  news coverage influenced by, 22
  slogans, 19
  tar content listing in, 67
  youth response to, 13, 23, 36, 44-
    45
African-Americans, tobacco use
  health risks, 75
  media depiction of, 22
  predictors of, 45-46, 48
  preferences, 75
  prevalence, 75
  weight gain with cessation, 134
  youth, 37, 46, 48
age
  of initiation of tobacco use, 35-36,
    37, 39, 49, 244
  and nicotine dependence, 87
  and tobacco use, 75, 143
Agency for Health Care Policy and
    Research (AHCPR)
  guidelines, 238-248
aggression, as abstinence effect, 214
airlines, smoking restrictions, 8-9,
    19, 153-154
akathesia, 119-120
alcohol use. *See also* substance use
  in dependent smokers, 116
  and tobacco use, 49, 50, 122-125
  and weight gain, 135
alcoholism treatment, 123
Alexander, Jennifer and Paul, 144
Alzheimer's disease
  nicotine and, 74, 76-79

visual threshold in, 70
American Heart Association, 160,
  238
American Psychiatric Association
  cessation guidelines, 238-239
  Diagnostic and Statistical Manual
    diagnosis, 239
American Psychological Association,
  51
American University, 218
analgesic effects, 11
Andreski, Patricia, 115-116
antagonist, nicotinic, 77, 111
antidepressant, cessation treatment
  using, 235, 243
anti-tobacco campaigns. *See also*
  tobacco control movement
  Arizona, 36
  Bears for Butts campaign, 113
  effectiveness, 8, 36, 44, 223
  nineteenth century, 14
  television messages, 20
anxiety. *See also* anxiolytic effects
  in abstinent smokers, 108, 214
  caffeine and, 122
  in dependent smokers, 116
  nicotine-related, 79
  and risk for smoking, 107
  in weight-control smokers, 137-
    138
anxiolytic effects, 10, 46, 47-48, 53,
  97
appetite suppression
  with phenylpropanolamine, 148
  smoking and, 48, 134, 136, 137,
    141-142, 211
Arcavi, Lidia, 141
Arendash, Gary W., 77-78
Arizona anti-smoking campaign, 36,
  170
Arizona State University, 49
Ark of the Covenant, 127

arousal
  dieting and, 146
  nicotine-related, 112-114, 122
Association for Public Health
    (London), 50-51
association, statistical, 42-43
asthma, environmental tobacco
    smoke and, 153, 157, 160, 163
athletes, smokeless tobacco use, 55
attention
  depression and, 69-70
  nicotine and, 69
Auburn University, 25-26
Audrain, Janet, 141
Australia
  chippers, 98
  National Health and Medical
    Council report, 160
  pregnant smokers, 179
  substance use by youth, 50
Austria, weight control in smokers,
    140
aversion strategies, as cessation
    treatment, 234, 244

B

Balbach, Edith, 23
Balfour, David J. K., 97
Bandura, Albert, 211-212
bans, smoking. *See* restrictions,
    smoking
Barendregt, Jan, 194-195
Barnum, Howard, 189
bars
  promoting cigarettes, 25
  restrictions on smoking in, 169
Bauman, Adrian, 179
Bauman, Karl E., 79-80
Bayard, Steven, 164
Bears for Butts campaign, 113
behavior

compensatory, 134, 135, 166, 167-
    168
desensitization, 78
functional base for, 188
linked, 120-121, 211-212, 215
reinforcement with nicotine, 185,
    190, 211-212, 215
stages of change theory, 26-27
values and, 188-191
behavioral economics of tobacco use,
    177-181, 183-187, 188, 190
Benet, Stephen Vincent, 86
Bernstein, Martine, 146-147
Bickel, Warren, 177-178, 179-181,
    183-185
biological markers of tobacco use,
    17, 71
Blacks. *See* African-Americans,
    tobacco use
Blaze-Temple, Debra, 50
Bolinder, Gunilla, 52, 54-55
Borneo, smokers in, 142, 144
Botvin, Gilbert, 45-46
brain
  chemistry, 95-97, 112
  lateralized function, 70, 111-112
  nicotine in the, 111
  physiology, 111-112
Breslau, Naomi, 50, 115-116
British Royal College of Physicians,
    223
Brown & Williamson, 41
bupropion, 235, 243, 248
Burns, David M., 193
butts, cigarette
  children's risk from, 167
  as litter, 167
  measuring intensity of smoking,
    166-167
buying of tobacco, by adolescents.
    See access to tobacco; sales,
    tobacco

C

caffeine
  addictiveness, 92-93
  and metabolic rate, 145
  nicotine compared to, 61, 92-93
  nicotine used with, 120-122, 145
  withdrawal, 220
California
  beach debris, 167
  Bears for Butts campaign, 113
  chippers, 98
  Environmental Protection Agency
    (state) report, 157-158
  prevalence of tobacco use, 18, 44,
    45, 167
  restrictions on smoking in, 163,
    169
  Smoke Free Cities, report, 162
  youth use of tobacco in, 44, 45
caloric intake, abstinence and, 139,
  214
Camel brand cigarettes
  history, 41
  marketing, 19, 20, 41
  nicotine in, 67
  promotions, 25
  youth preferences for, 23, 45
Camel, Joe. *See* Joe Camel
Camp, Diane, 48
campaigns. *See* anti-tobacco
  campaigns
Canada
  cigarette taxation in, 7-8
  Health and Welfare, 91
  Non-Smokers' Rights Association,
    236
  Royal Society, 91
cancer, tobacco-related. *See also* lung
  cancer, tobacco-related
  African-Americans, 75
  cigar use and, 198, 199

  from environmental tobacco
    smoke, 157
  filters and, 9
  in India, 14
  mortality risk, 75
  smokeless tobacco, 56
  Whites, 75
candy cigarettes, 43
carbon monoxide, as indication of
  smoking, 17, 71
carboxyhemoglobin, 71
Cardador, Teresa, 26
cardiovascular disease
  in African-American smokers, 75
  cigars and, 199
  environmental tobacco smoke and,
    156-157, 159-160, 163
  reduction of tobacco use and, 192,
    234
  and relapse of smoking, 99
  risk in smokers, 16, 18, 75, 205
  smokeless tobacco and, 18
  and weight loss with cessation, 135
  in white smokers, 75
cardiovascular system, nicotine stress
  on, 18
Caribbean, tobacco use, 12
Carmelli, Dorit, 122, 124, 135
Carson, Rachel, viii
Caskey, Nicholas, 120
Cat's Eye, 249
Centers for Disease Control and
  Prevention, 37, 40, 50, 51, 133-
  134, 140, 200
cervical cancer, 158
cessation of tobacco use. *See also*
  cessation treatment, tobacco;
  relapse; weight gain, cessation-
  related
  alcoholism treatment and, 123,
    124
  attitudes of current smokers, 233

caffeine and, 120
cold turkey, 204, 236-237
compensatory behavior, 134, 135, 167-168
cutting down, 168, 192
depression and, 116
expectations about success in, 218-219
gender differences in, 146
gimmicks, 231-232
guidelines for, 238-249
health benefits, 18-19
and health care costs, 194-195
ideal candidate for, 217-218
lapse, 209-210
learning to quit, 249
self-quitters, 236-237
research problems, 219
"stages of change" theory, 26-27
success predictors, 49-50, 217-218
temporal perspective of, 181, 183
youth, 39
cessation treatment, tobacco. *See also* cessation of tobacco use; nicotine replacement treatment
access to, 8, 216, 223
antidepressants in, 235
aversion strategies, 210, 231, 233, 234
behavioral economics and, 179-180
brand switching, 233
cognitive-behavioral, 147
community-based efforts, 192, 205-209
counseling, 243, 248
devices for, 231-232
exercise with, 148
gender differences in, 145
harm reduction strategies, 191-193
health care system responsibilities, 245-248

hypnosis, 232
ideal, in economic perspective, 184
intensive, 243-244
nicotine fading, 233
pharmacological, 148, 210, 222-223, 235-236
for psychiatric patients, 125-126
treatment matching in, 243
weight control with, 138-139, 140, 148
Charlton, Ann, 38
Chassin, Laurie, 49
Chen, Kevin, 49
children. *See also* youth, tobacco use
health risks from environmental tobacco smoke, 157, 160, 163
China, smoking in
mortality projections, 17, 38
prevalence, 14, 15
tobacco production, 15
chippers, tobacco, 95, 98
Cigar Aficionado, 27, 113, 198
cigarettes. *See also* generic brands of cigarettes; rates, tobacco use
compared with other substance use, 219-220
cost of, 12, 178
health costs per pack, 182
lettuce, 232
low-nicotine, 193, 233
machine-rolled, 12, 13
pack, number in, 18
sales in U.S., 182
smoked daily in U.S., 18
smokeless, 191-192
cigars
consumption in U.S., 198
costs of, 195
environmental smoke from, 196
famous smokers, 195, 198
health risks from, 198, 197-200, 233-234

disease concept, 97-99
female smokers, 92, 141-142
measuring, 91-92
smokeless tobacco, 52, 53-53
tolerance and, 90, 95
depression
in abstinent smokers, 108, 166,
214, 221
and cessation of tobacco use, 116,
235, 247
EEG activity, 114
gender differences in, 141-142
nicotine and, 79, 114-118, 142
prevalence and description, 110,
114-115
and smokeless tobacco, 53
and smoking, 46-47, 48, 107, 110,
115-118
in weight-control smokers, 137-
138
in youth, 46-47, 48
desensitization
behavioral, 78
of nicotinic receptors, 97
Detroit, youth access to tobacco in,
40
developing countries
smoking in, 14, 38
smokeless tobacco use, 54
diagnosis, psychiatric, 239
Diagnostic and Statistical Manual
diagnosis, 239
DiClementi, C. C., 26
dietary patterns
cessation of tobacco use and, 140-
141, 248
restraint, 139
and success in quitting, 145-146
tobacco use and, 124
Dilbert cartoon, 113
disease, nicotine dependence as, 97-
99

disease, tobacco-related. *See*
morbidity, tobacco-related
Dole, Bob, 85-86
dopamine, 96
dose-response relationship, 109, 111
down regulation of receptors, 96
driving simulation, nicotine-related
performance, 68
drug abuse. *See* substance use
drug-liking scales, nicotine rated on,
109
drugs, illegal. *See* substance use
Duke University, 62, 76
dysthymia, 115

E

eating disorders, 73-74, 107
economics, tobacco. *See also*
behavioral economics; health
care costs, tobacco-related
cost of smoking, 12, 182, 193-195
expenditures for tobacco products,
13
public health policy and, 195
value added from ton of tobacco,
189
education about cessation, for
clinicians, 241
educational attainment
and cessation success, 49
and weight gain, 146
EEG. *See* electroencephalography
efficacy, feelings of, 46
Eisner, J. R., 47
elastic supply/demand, tobacco, 185-
186
electroencephalography
of brain response to nicotine, 112
of depressed smokers, 114
event-related potentials, 65
of personality indicators, 114

Elrod, Karey, 62
emphysema, tobacco-related, 16
employment, and cessation of
    smoking, 49
Environmental Protection Agency,
    U.S.
  report on environmental tobacco
    smoke, 157, 162-163, 164-165
environmental tobacco smoke
  carcinogens in, 71, 158
  and children's performance, 79-80
  chronic effects, 155-157
  from cigars, 196
  economic costs per pack, 182
  definition and description of, 155
  discomforts from, 154, 157
  health risks of, 16, 154, 156
  immediate effects, 155
  lawsuits, 170
  mortality from, 16, 157
  in psychiatric units, 125
  reports about, 158-164
  smokers' rights literature, 26
  Surgeon General's report on, 157
  and weight gain, 147
Erasmus University, 194
Erickson, Michael, 200
esophageal cancer, 158
euphoriant, nicotine as, 10, 92, 93,
    109, 126
Europeans, tobacco use, 11, 38, 52
exercise. *See* physical activity
experimentation, tobacco, by youth.
    See initiation of tobacco use

F

Fantasticks, The, 36
females, tobacco use. *See also* women
  African-Americans, 38
  cessation, 218, 246
  in China, 12

cigar marketing to, 199
dependence on nicotine, 141-142
and depression, 48, 117, 141-142
in developing countries, 14
gender roles and, 13
in Indonesia, 14
initiation age, 39, 142
media depiction of, 22
metabolism in, 145
response to nicotine, 144-146
risk factors, 142
smokeless tobacco, 55
in Spain, 12
weight control and, 48, 135-138
youth, 37, 44, 48
Fiji, smoking cessation in village,
    205-208
films, smoking depicted in, 20-22
filters, cigarette, 9
Fiore, Michael, 238, 241
fires, smoking-related, 182
Fisher, Paul M., 23
fixation on smoking, in abstinent
    smokers, 73-74
Flegal, Katherine M., 140
Food and Drug Administration, U.S.
  control of tobacco, 100
  nicotine regulation, 15
Francis Scott Key Medical Center,
    smoking restrictions at, 165-168
Freedman, Robert, 119
frequency of tobacco use, and
    persistence of use, 49
Frost, Karen, 22

G

Gammon, Marilie, 158
gender differences
  alcohol-nicotine effects, 122, 124
  in cessation success, 218
  in metabolism, 145

in nicotine effects, 144
in rates of tobacco use, 12, 14, 142-144
in sensitivity to nicotine, 144, 218
weight gain with cessation, 134, 136
generic brands of cigarettes
expense of, 7
motivation for choosing, 23
tar and nicotine information, 67
genetic factor, nicotine, alcohol, and coffee consumption, 124
Gilbert, David G., 66, 70, 114
Gilpin, Elizabeth, 49-50
Glantz, Stanton, 20, 21, 26
Glover, Elbert, 54
Gold, Mark, 8
Gori, Gio B., 164
Gossop, Michael, 219, 221
Greene, Jody, 38
Gritz, Ellen, 136
Gross, Todd M., 73
Groth-Marnat, Gary, 208
group self-identification, 46
Grunberg, Neil E., 134, 209
gum, nicotine, 64, 66, 69, 76-77, 148, 191, 192, 234-236, 237. *See also* nicotine replacement treatment

H

Haaga, David A. F., 218
habit, addiction contrasted with, 8, 92-95
Hall, Sharon, 117-118, 140, 148
hard drugs. *See also* substance use
nicotine compared with, 219-220
harm avoidance, 47
harm reduction strategies, 180-181, 191-193
Harvard University, 160

Hatsukami, Dorothy, 53
Hawaii, health rating, 18
Hazan, Anna, 20, 21, 26
hazard, tobacco-related. *See* risk, tobacco-related
Health and Welfare Canada, 91-92
health care costs, tobacco-related, 182, 189, 193-195
health care system, responsibilities for cessation, 245-248
health effects, tobacco-related. *See* risk, tobacco-related
heart attacks. *See* cardiovascular disease
heart rate, abstinence and, 214
Heishman, Stephen J., 63-64
Henningfield, Jack, 7-8, 89, 191-192, 196-197
Henry Ford Health Sciences Center, 50
history, tobacco
native tribal use, 10-11
origins and spread, 10-13
smokeless, 52
homeless youth, tobacco use, 38
Hôpital de la Salpêtrière (Paris), 64
hospitalized patients, smoking cessation, 247
hospitals
restrictions on smoking, 19, 165-168, 245, 247
smoking in psychiatric units, 125-126
Howard, George, 159-160
Howard University, 74
Hughes, John, 219-220
hunger, as abstinence effect, 214
Hursh, Steven R., 185-186
Hurt, Richard, 125, 235
hyperactivity, 107
hypnosis, as cessation treatment, 232

Le Houezek, Jacques, 64-65
learning disabilities, and tobacco use, 38, 78, 79
legislation, tobacco, 14-15, 25
LeGuin, Ursula, 69-70
Leistikow, Bruce, 117
Leonard, Sherry, 119
Leslie, Simon, 208
Levin, Edward D., 62
Licari, Michael J., 191
life expectancy, tobacco use and, 194, 244
Life Skills Training program, 51
Lipton, Helene L., 21
Lo, Sing Kai, 50
London Institute of Psychiatry, 76
longitudinal research, 42, 46, 48-49
loss of control phenomenon, 184
lung cancer, tobacco-related
  cigars and, 199
  environmental tobacco smoke and, 156, 160, 162, 163, 164-165
  mortality, 18, 143
  and relapse of smoking, 99
  smoking-related, 16, 18
  women, 143

M

Maag, John W., 38
Mac the Moose, 41
machine-rolled cigarettes, 12, 13
Maine American Cancer Society, 41
mainstream smoke, 155. *See also* environmental tobacco smoke
marketing, tobacco, 14, 25-26, 199
Marlatt, George, 216
Marlboro cigarettes, 44, 46, 67
Marlboro man, 8, 19-20
marriage, and cessation of smoking, 49

Massachusetts
  cigar smoking by youth, 199
  Web site for nicotine levels, 67
Massachusetts General Hospital, 186
Mâsse, Louise, 47
Mayo Clinic, 125
McKenna, F. P., 224
mecamylamine, 77, 111
media portrayals of tobacco use, 6, 20-22
Medical College of Georgia, 62
Meier, Kenneth J., 191
memory, nicotine and, 69, 72, 77-78
Memphis State University, 48
menstruation, and cognitive capacity in abstinent smokers, 73
mentholated cigarettes, 9, 75
metabolic rate
  as abstinence effect, 214
  and activity, 145
  cessation and, 135-136, 140-141
  drug, as abstinence effect, 214
  gender differences in, 145
  nicotine and, 140-141
  and weight gain, 135-136, 140-141
methodological problems
  naturalistic/realistic, 41-43
  never-smokers compared to ex-smokers, 65-66
  generalizability of results, 66, 68, 121, 221
Meyers, Andrew W., 138
Middle Tennessee State University, 224
Midwestern Prevention Project, 51
Miller Matthew, 193
Miller, Norman, 99
mood. *See also* depression
  dieting and, 146
  nicotine and, 10, 66, 70, 96, 108-114, 117

morbidity, tobacco-related. *See also* cancer, tobacco-related; cardiovascular disease; risk, tobacco-related
on Indian subcontinent, 14
worldwide, 5-6
Mormons, 44
Morse, Rob, 169
mortality, tobacco-related. *See also* risk, tobacco-related
age of initiation and, 244
in developed countries, 244
economic costs of, 182, 189
from environmental tobacco smoke, 16, 157
per ton of tobacco consumed, 189
predicted, 14, 17, 37, 39, 40, 234
premature, likelihood of, 178
time lag from tobacco processing to, 189
worldwide, vii, 5-6, 13, 17, 244
Mr. Dip Lip, 231
Mr. Gross Mouth, 231
Murray, Bridget, 51

N

National Academy of Sciences-National Research Council Twin Registry, 135
National Cancer Institute, 199
National Center for Health Statistics, 133-134
National Household Survey on Drug Use, 91-92
National Institute on Drug Abuse, Addiction Research Center, 63, 66
National Smokers Alliance, 163, 170
National Working Conference on Smoking Relapse, 209
Native Americans, tobacco use prevalence, historical and current, 10-11

in sacred ceremonies, 27-28
smokeless tobacco use, 53, 55
nausea, nicotine-related, 10, 65, 94, 155
negative emotions, tobacco use and, 117-118, 218
Nelson, Donald, 200
neuroticism, 116
neurotransmitters, 95-96, 112
Nevada, youth use of tobacco, 44
New Hampshire, youth access to tobacco in, 40
New York University, 77
Newhouse, Paul A., 77
news reports, tobacco-related
health risks, 20
in *Scholastic News*, 23
smoking mentioned in, 18
in *Weekly Reader*, 23
Nicotiana, 9-11
nicotine. *See also* cessation, tobacco; dependence, nicotine; nicotine replacement treatment
and Alzheimer's disease, 76-79
antagonist, 77, 111
biophysical effects, 10
caffeine used with, 120-122, 145
in cigars, 196-197, 198
content information, 67
dose-response relationship, 109
fading, 213, 233
fetal effects of, 161
gum, 64, 66, 69, 76-77, 234-236
interactions with psychiatric medications, 108, 120, 126
levels in tobacco products, 67
limiting content, 193
manipulation in tobacco products, 64, 100-101
metabolic effects, 140-141
patch, 76, 78, 234, 237
poisoning, 167

Pickworth, Wallace, 89
Pierce, John P., 44-45, 49-50, 142
pipe smoking, 11
pleasure, smoking-related, 47, 109
Pomerleau, Cynthia, 109, 126, 136-137
Pomerleau, Ovide, 109, 126, 222
popularity of tobacco, 8-9
poverty levels, smokers below, 178, 223
prediction, statistical, 42-43
preference reversal, 184
pregnancy
  counseling during, 246
  and depression, 115
  fetal effects of nicotine, 161
  tobacco use during, 62, 161, 178-179
prenatal nicotine exposure, 62-63, 161
prevalence, tobacco use. *See* rates, tobacco use
prevention of tobacco use
  effectiveness on youth, 36, 47, 50-51
  Mac the Moose, 41
  timing, 50
print media, tobacco risk representation, 22-23
Pritchard, Walter, 93-94
Prochaska, J. O., 26
productivity losses, 182, 193
promotions, tobacco
  expenditures on, 23, 25, 45
  youth targeted by, 36, 44-45
psychiatric patients
  cessation treatment for, 125-126, 239
  rate of smoking, 107
psychic dependence, 90
psychoactive effects, 69, 193

psychophathology, cessation of tobacco use and, 118
psychosocial scars, 46
public health policy, economics and, 195
purchasing of tobacco. *See* access to tobacco; sales, tobacco
Pure Food and Drug Act of 1886, 15

R

Rachlin, Howard, 188-190
Raeburn, Paul, 158
Rásky, Éva, 140
rates, tobacco use
  African-Americans, 75
  age and, 75
  Canada, 8
  cigarette smoking, 18, 39, 51, 75, 107
  cigars, 37, 197-199
  developing countries, 14
  foreign, 12, 14-13
  gender differences, 12, 14, 142-144, 146
  physicians, 12
  pregnant women, 161
  schizophrenia and, 107
  smokeless tobacco, 8, 37, 51, 53, 55
  United States, 13, 18, 44, 45, 167, 168, 191
  whites, 75
  worldwide, vii, 12, 13, 222
  by youth, viii-ix, 7, 12, 36-37, 39, 51
reactions, enhanced by nicotine, 70
receptors, nicotinic, 97
Red Kamel cigarettes, 41. *See also* Camel brand cigarettes
Redhead, C. Stephen, 162
Reid, Donald J., 50-51

reinforcing effect of nicotine as, 185, 190, 211-212, 218
relapse
behavioral economics and, 189-190
cardiovascular disease and, 99
culture and, 206, 203
description, 209-211
emotional state and, 117, 146
expectations of, 218-219
gender differences in, 218
lung cancer and, 110
nicotine reinforcement and, 211-212
risk factors for, 48-49, 204, 209-211, 215-216, 218-219, 221-222, 243, 249
time to, 233
weight gain and, 139, 148
women, 145-146
relaxation, tobacco use and, 53, 10
religion
and cessation of smoking, 206-208
and smoking, 13, 28, 44
Renneker, Mark, 208
research methods. *See also* methodological problems
animal studies, 62, 77-78, 94
cross-sectional studies, 40-43
double-blind conditions, 65
longitudinal studies, 42, 46, 48-49
measurement of tobacco use, 17-18, 166-167
recruitment of subjects, 66
Research Triangle Institute, 38
Resnick, Michael, 125
respiratory illness
in children, 155, 160
environmental tobacco smoke and, 155, 160, 164-165
smoking and, 157
Restak, Richard, 95-96

resting energy expenditure. *See* metabolic rate
restlessness, as abstinence effect, 214
restrictions, smoking
airlines, 8-9, 19, 26, 153-154
Arizona, 170
California bars, 169
effectiveness, 8, 166, 168, 191
hospitals, 19, 165-168
prisons, 170
public buildings, 19, 26
restaurants, 19, 26, 169, 170
study of, 165-168
Surgeon General's report and, 19
workplace, 9, 19, 26, 163, 165-166, 168
Reynolds, Richard, 41
rights, smokers'. *See* smokers' rights
Rigotti, Nancy, 186-187
risk, tobacco-related. *See also* morbidity, tobacco-related; mortality, tobacco-related
awareness of, 51, 179, 223
cancer mortality, 75
cessation and, 239
debate, 15
media representation of, 20, 22-23
perceptions of, 53, 191, 224
smokeless tobacco, 51, 53, 54-55
style of smoking and, 193, 196, 197, 199
threshold approach, 162
weight concerns relative to, 137, 147
weight-of-evidence approach, 157
risk-taking, and tobacco use, 53, 69, 171
RJR Nabisco. *See also* Camel brand cigarettes
Weekly Reader ownership, 23
Robert Wood Johnson Foundation, 197

Robinson, John, 93-94
Rohde, Paul, 46-47
Roswell Park Cancer Institute, 197
Rotfeld, Herb, 25-26
Rowberg, Richard E., 162
Royal Society of Canada, 91-92
rural environments, tobacco use, 37-38
rush, nicotine, 109, 122, 144
Russell, Michael, 97, 99
Russia, tobacco use in, 12, 14

S

sales, tobacco. *See also* access to tobacco
  bans on, 14
  cigars, 7, 198
  to minors, 40, 186-187
  smokeless tobacco, 55
  vending machines, 25
Samuelson, Robert J., 165
San Francisco Veterans Affairs Medical Center, 140
Sarvela, Paul, 37-38
schizophrenia
  in abstinent smokers, 108
  described, 118-119
  and nicotine, 118-120
  and smoking prevalence, 118, 119
Scholastic News, 23
Science Applications International Corporation, 185
Scruton, Roger, 6
secondhand smoke. *See* environmental tobacco smoke
self-control, 188, 190
self-esteem, 46
self-quitters, 236-237
Seventh Day Adventists, 44
sex differences. *See* females, tobacco use; gender differences

sexual behavior, risky, and smoking, 50
Sherwood, Neil, 68-69
Shiffman, Saul, 94
Shumaker, Sally A., 209
sidestream smoke, 155, 196. *See also* environmental tobacco smoke
Silent Spring, viii
Skinner, B. F., 211-212
slaves, tobacco traded for, 11
sleep disturbance
  in abstinent smokers, 214
  nicotine therapy and, 79
smokeless tobacco use
  cardiovascular effects of, 18
  cessation, 248
  dependence on, 52
  factors associated with use, 8, 53
  as harm reduction strategy, 191, 233-234
  health risk, 51-52, 54-55, 233-234
  history, 52
  initiation of, 53, 55
  perceptions of risk, 53
  rates, 37, 51, 53, 55
  restrictions on, 52
  sales revenues, 55
  youth, 8, 37, 38, 44-45, 51, 197
smokers' rights
  limits on, 154, 165
  literature, 26-27
  Weekly Reader promotion of, 23
smoking, cigarette. *See also* cigarettes; tobacco
  bans, *see* restrictions, smoking
  cultural influences, 8
  demographics shift among users, 222
  depicted in films, television, 18-20
  depicted in news, 18-21
  in developing countries, 14